T0296136

ATOMIC SPECTRA

In two volumes

VOLUME I

ATOMIC SPECTRA
AND THE VECTOR MODEL

BY

A. C. CANDLER

Sometime Scholar of Trinity College,
Cambridge

'There is one thing I would be glad to ask you. When a mathematician engaged in investigating physical actions and results has arrived at his own conclusions, may they not be expressed in common language as fully, clearly and definitely as in mathematical formulae? If so, would it not be a great boon to such as we to express them so—translating them out of their hieroglyphics that we also might work upon them by experiment.'

Letter from MICHAEL FARADAY to CLERK MAXWELL

VOLUME I

SERIES SPECTRA

CAMBRIDGE
AT THE UNIVERSITY PRESS
1937

CAMBRIDGE
UNIVERSITY PRESS

University Printing House, Cambridge CB2 8BS, United Kingdom

Cambridge University Press is part of the University of Cambridge.

It furthers the University's mission by disseminating knowledge in the pursuit of education, learning and research at the highest international levels of excellence.

www.cambridge.org
Information on this title: www.cambridge.org/9781107505803

First published 1937
First paperback edition 2015

A catalogue record for this publication is available from the British Library

ISBN 978-1-107-50580-3 Paperback

PREFACE

The empirical laws on which modern spectroscopy is based were all worked out with a notation which seems to-day cumbrous and misleading; yet since 1929, when a large group of physicists adopted the modern simplifications as standard, no adequate review of the earlier work seems to have appeared in this country; the beginner must still refer to Fowler's *Report* and master a notation which he will soon wish to forget. To repair this omission is the chief aim of the first volume. In addition, an account is given of the splitting of spectral lines in a magnetic field, since this is essential to a full understanding of the series laws.

As Fowler's *Report* ordered series spectra, so Hund's *Linienspektren* ordered the spectra of the elements and related them to the periodic system; if the evidence was impressive in 1927, it is now overwhelming; the theory is developed and the spectral types described in five chapters of the second volume.

Besides these two chief topics, there are others which are more or less closely related; some of these, notably hyperfine structure, are of very recent growth. The last chapters take up these topics one by one, and lead the reader to the boundaries on which research is now concentrated.

In the hope that the argument may remain within the easy comprehension of every working physicist, the vector model has been used throughout. Where the quantum mechanics has refined the predictions of the model, the results have been freely quoted and compared with experiment; but the rather heavy mathematics, which is the basis of even the simplest calculation, has been wholly excluded.

It is a pleasure here to record how much this book owes to the Cambridge Professor of Mathematical Physics, R. H. Fowler; without his encouragement it would never have been begun, or if begun, it would without his interest never have seen the light of day. To many others I am indebted for help in the later stages; J. E. Keyston was kind enough to write a draft chapter on hyperfine structure; Dr R. W. B. Pearse has read chapters x, xvi and xxii, and advised me on them.

So many have helped me with the illustrations that I can here do little more than record their names and my gratitude. Professor H. Dingle has obtained me the loan of many photographs, which are the property some of Professor A. Fowler and some of the Imperial College; he has also advised me on many other points. Professor E. Back has sent me the photographs from which the plates in his book on the 'Zeeman-effekt' were made. I am indebted to Professor C. J. Bakker for some photometer curves of hyperfine structure, to Mrs K. Darwin for drawings of the Paschen-Back effect, to Dr R. G. J. Fraser for permission to copy deflection patterns from his book on Molecular Rays, to Professor J. B. Green for photographs of the incipient Paschen-Back effect and for blocks illustrating hyperfine structure, to Dr D. A. Jackson for photographs of hyperfine structure, to Professor W. F. Meggers for photographs of scandium multiplets, to Professor F. H. Spedding for absorption spectra of samarium and finally to Professor R. Tomaschek for the spectra of various phosphors.

The editors and publishers of many periodicals have kindly given me permission to copy from their works. Of these debts I hope the detailed references given beneath the figures are the best acknowledgment.

In a work in which so much detail appears, one cannot expect that errors have been entirely avoided; in the hope that these are not numerous, I shall be glad to know where they occur.

A. C. C.

LEIGHTON PARK
READING

CONTENTS

VOLUME I. SERIES SPECTRA

LIST OF PLATES

CHAPTER I
INTRODUCTION

1. The problem

When an atom is heated in a flame or excited by a spark discharge, it emits an immense number of lines, some perhaps visible but many more in the ultra-violet; to reduce these lines to order and to correlate them with the structure of the atom have been the two ideals towards which spectroscopists have always worked. The problem has been tackled both inductively and deductively; the observed spectrum has been arranged in systems and series, and the states of the atom have been deduced; and from the other end the states of the atom have been calculated with the aid of the simple vector model or the more elaborate wave mechanics, and the lines deduced have been compared with experiment. History shows that advance has been most rapid when the two methods have been used side by side; but in reviewing the work of the past, a better grip of the subject is probably obtained by presenting experiment first, and introducing theory only as a convenient method of correlating the facts; this order has at least the advantage that should the present theory ever be displaced by a better, the experimental facts would still be intelligible.

Half a century ago this work was scarcely begun, yet to-day one may simply, if also only roughly, predict the spectrum of any nuclear charge with any number of extra-nuclear electrons. Half the elements have been analysed in some detail, and there can be no doubt that in due course every line of every element will be labelled as arising from a particular transition within the atom. The scheme has already been developed; what remains will in all probability reveal no unsuspected difficulty, but will be a long piece of routine work.

2. Historical

The history of the investigation of spectra may be divided into four periods, an acoustic period, a series period, a quantum-theory period and a last period introduced by the quantum mechanics of

Schrödinger, Heisenberg and others. Like most other historical divisions, these are arbitrary; the periods really overlap and inter-weave, but it is necessary to be arbitrary in order to be clear.

The first period began with the earliest measurements of wave-length, and continued with the work of Boltzmann, Liveing and Dewar until 1881, when Schuster brought it to an abrupt conclusion. During this period any theories put forward were based on analogies with the harmonic ratios of sound. The one triumph was Johnstone Stoney's discovery that in the spectrum of hydrogen the frequencies of three of the four visible lines are in the ratio of 20:27:32.* Schuster, however, suspected that this line of attack was unprofitable, and in a very interesting paper justified his suspicion by proving that the closeness of fit was only what chance would predict.†

Though he had stopped unprofitable speculation, Schuster put forward no new theory, and the second period did not open until four years later, when Balmer published his classic formula giving the wave-lengths of the visible lines of hydrogen. Written in a modern form, this states that the wave-length λ and the wave-number ν are given by the equation

$$\frac{1}{\lambda}=\nu=R\left(\frac{1}{2^2}-\frac{1}{n^2}\right).$$

R is here an absolute constant now commonly known as Rydberg's constant, while n assumes the values 3, 4, 5, 6 for the four visible lines.

This formula served as a first example of two important generalisations. Rydberg in the last decade of the century showed how the simple Balmer formula might be generalised to give an account of series in other elements. While later Ritz pointed out that written in the form

$$\nu=\frac{R}{n_1{}^2}-\frac{R}{n_2{}^2},$$

where n_1 and n_2 are both integers, the Balmer formula states that

* Johnstone Stoney, *PM*, 1871, **41** 291. A key to the letters used in referring to periodicals is given in Appendix I.

† Schuster, *PRS*, 1881, **31** 337.

the wave-number of any spectral line may be written as the difference of two terms T' and T'',

$$\nu = T'' - T'.$$

This is the important Rydberg-Ritz combination principle; though stated in 1908 before Bohr applied the quantum theory to solve atomic problems, it is so closely linked with his ideas that it is best considered as the first dawn of the third period.

In 1913 Bohr applied Planck's quantum theory to solve the problem of the radiating atom. He postulated that an electron, which revolves round the nucleus, must move in one of a number of orbits specified by quantum conditions; moving in one of these 'stationary states' the electron does not radiate as classical electrodynamics requires; instead radiation is emitted only when the electron jumps from one stationary state to another. In a second postulate Bohr added that the frequency of the line emitted depends on the difference in energy of the two states.

On this basis the whole complex structure of modern spectroscopy has been erected. Its great achievement is that it enables physicists to deduce from the spectrum of an element the energies of the stationary states of the atom. To correlate the energies of these states into some system was the next problem; a first step had been made in the work on series of an earlier day, and this has been continued in the so-called 'vector model' of the last decade. Conceived to describe the atom in terms of the classical mechanics and the theory of relativity, the model is to-day regarded simply as a convenient way of correlating experimental facts.

Like the geometrical theory of optics in another field, the vector model gives a satisfactory description of many spectroscopic laws. It is simple and easily visualised, and for these reasons it is valuable. But it covers only a part of the area covered by the wave mechanics, just as in optics the geometrical theory is less powerful than the wave theory.

The fundamental equations of the wave mechanics offer in principle a solution of any problem of atomic structure, and so also of spectral emission; but in practice the equations are so difficult to solve that mathematicians have been able to apply the new theory only to some of the simpler problems.

The advent of the wave mechanics heralded the dawn of what must undoubtedly be considered a fourth period in the development of spectroscopy; but important as the new theory is, it has not, and almost certainly will not, displace the older vector model, for the physicist asks for something he can touch and see, or, as he often says, for something he can understand. This the vector model gives him, and with it the bulk of this book is concerned. Many of the results obtained by the wave mechanics will be noted, but the theory which underlies them will not be discussed.

CHAPTER II

THE HYDROGEN ATOM

1. Series

The hydrogen atom is the only neutral atom which does not contain more than the two bodies with which alone classical mechanics is able to deal, and if the band spectrum due to the molecule be ignored the spectrum is also simple. The lines produced by the atom are shown in Fig. 2·1, and may be divided into five groups, which are called technically 'series'.

Series occur not only in hydrogen, but in all simple spectra; starting with the line of longest wave-length in any series and moving towards the violet, one observes that the lines lie nearer and nearer together and at the same time decrease in intensity. When a large number of lines are visible, it is quite clear that the decreasing separation at the violet end of the series piles up the lines towards a sharply defined limit.

Experience shows that spectral laws take their simplest form when expressed as relations between frequencies rather than wave-lengths, and therefore in Fig. 2·1 as in all subsequent figures the scale is a uniform scale of frequency. But in that spectroscopists measure the wave-length and can calculate the absolute frequency only when they know the velocity of light, they commonly use the wave-number or reciprocal of the wave-length rather than the frequency. The visible spectrum extends roughly from 4000 to 8000 A., when 1 A. or angstrom is 10^{-8} cm. or $10^{-1}\mu\mu$. In terms of wave-numbers this is 25000 to 12500 cm.$^{-1}$ Again small separations are often important and it is sometimes convenient to remember that the separation of the yellow $D_1 D_2$ doublet of sodium is 6 A. or 17 cm.$^{-1}$

The five series of the hydrogen spectrum are named after their discoverers; of these Balmer* was the first, for he showed, as early as 1885, that the four lines in the visible region may be

* Balmer, *AP*, 1885, **25** 80.

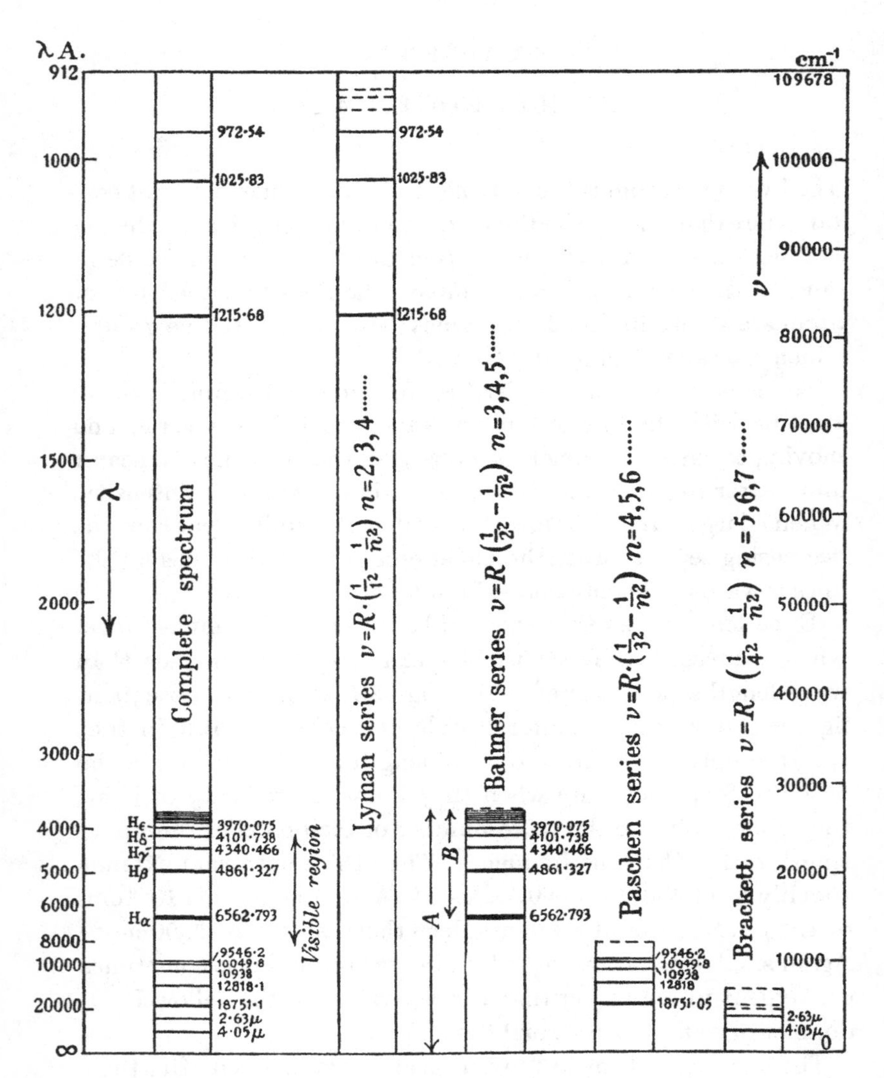

Fig. 2·1. Spectrum of hydrogen. After Grotrian, *Graphische Darstellung der Spektren.*

represented by a formula, which in modern notation takes the form

$$\nu = R\left\{\frac{1}{2^2} - \frac{1}{n^2}\right\}, \qquad \text{......(2·1)}$$

where R is a constant of value 109677·76 cm.$^{-1}$, and n takes on the successive values 3, 4, 5, 6 for the four lines. Later workers measured many lines in the ultra-violet and found that these fit the formula with remarkable accuracy when n is given higher integral values. Fig. 2·2 gives a comparison of theory and experiment.

n	ν calculated[a]	λ calculated	λ observed[b]
3	15233·156	6564·628	6564·602
4	20564·760	4862·687	4862·680
5	23032·531	4341·685	4341·683
6	24373·049	4102·893	4102·891
7	25181·339	3971·195	3971·194
8	25705·950	3890·150	3890·149

Fig. 2·2. The first six lines of the Balmer series.

[a] R is taken as 109678·72, this being the value which Curtis found to give the best fit.

[b] Curtis, *PRS*, 1919, **96** 147. The wave-lengths have been corrected to vacuo.

Later Lyman* found another series of hydrogen lines in the far ultra-violet, while Paschen,† Brackett‡ and Pfund§ observed new series in the infra-red.

Generalised in the form

$$\nu = R\left\{\frac{1}{n_1^2} - \frac{1}{n_2^2}\right\}, \qquad \text{......(2·2)}$$

the formula showed itself capable of accounting for all these new lines; indeed it was used to predict some of them. n_1 is fixed in any given series and has the values 1, 3, 4, 5 for the Lyman, Paschen, Brackett and Pfund series respectively, while n_2 takes on a series of integral values subject to the condition that $n_2 > n_1$. The limit of any series is obtained by making n_2 tend to infinity, so that the limit of the Balmer series is simply $R/4$ or 27419·4 cm.$^{-1}$

* Lyman, *AJ*, 1906, **23** 181.

† Paschen, *AP*, 1908, **27** 537.

‡ Brackett, *AJ*, 1922, **56** 154; Poetker, *PR*, 1927, **30** 418.

§ Pfund, *JOSA*, 1924, **9** 193.

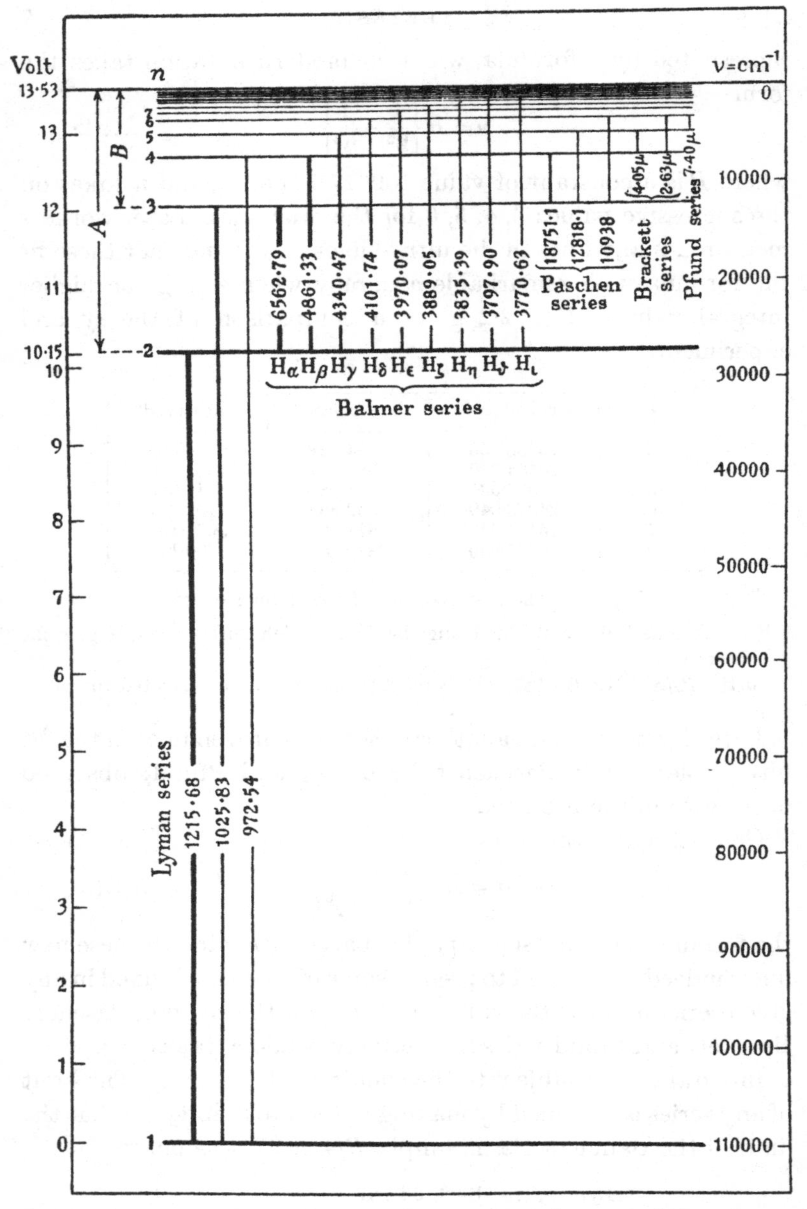

Fig. 2·3. Level diagram of hydrogen. Each vertical shows an electron jump; the numbers give the wave-length in angstroms, the thickness gives a rough indication of the intensity. After Grotrian, *Graphische Darstellung der Spektren*.

2. Interpretation by diagram

These facts are very simply illustrated by a diagram due to Grotrian (Fig. 2·3). Down a vertical line lengths R, $R/4$, ... R/n^2 are measured off, and horizontal lines are drawn through the

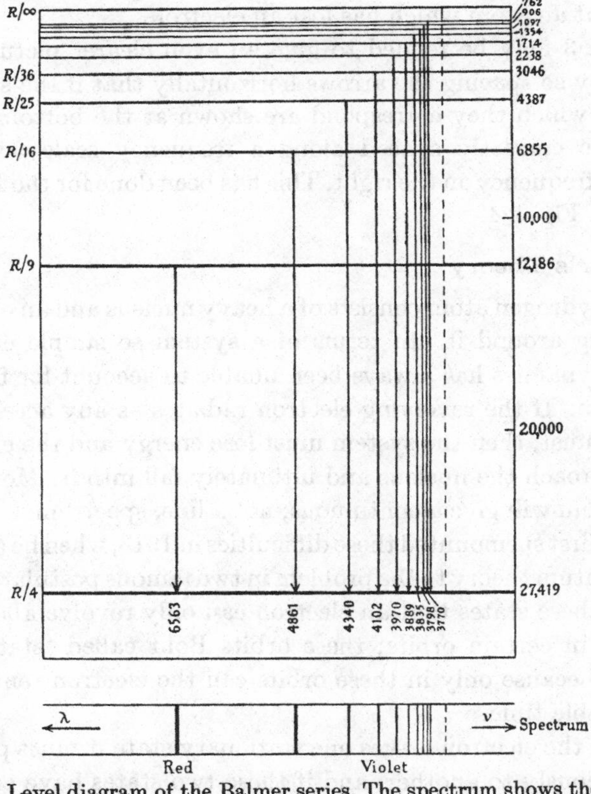

Fig. 2·4. Level diagram of the Balmer series. The spectrum shows the lines on a uniform scale of frequency.

points obtained to represent terms of the spectrum. Any transition between two terms may be represented by a vertical arrow, whose length will be a measure of the wave-number of the line emitted. The H_α line represented by the equation

$$\nu = R\left\{\frac{1}{2^2} - \frac{1}{3^2}\right\}$$

may be used to illustrate this; in both Figs. 2·1 and 2·3 it appears as the difference of two lengths marked A and B, which are in fact the quantities $R/4$ and $R/9$; but whereas in Fig. 2·1 they are measured from the limit of the Balmer series, in Fig. 2·3 they are measured down from a line which is the limit of the term sequence, being the value of R/n^2 when $n \to \infty$. The limit will be seen later to represent an atom which has lost an electron.

Fig. 2·3 may be refined to give an even clearer picture of a series, by so spacing the arrows horizontally that if the spectral lines to which they correspond are shown at the bottom, these lines are correctly spaced along a frequency scale with the highest frequency on the right. This has been done for the Balmer series in Fig. 2·4.

3. Bohr's Theory

The hydrogen atom consists of a heavy nucleus and an electron revolving around it. In terms of a system so simple classical electrodynamics has always been unable to account for the line spectrum. If the revolving electron radiates as any accelerated charge must, then the system must lose energy and the electron will approach the nucleus and ultimately fall into it. Moreover, the system will give a continuous, not a line, spectrum.

Bohr first surmounted these difficulties in 1913, when he applied the quantum theory to the problem in two famous postulates. The first of these states that an electron can only revolve about the nucleus in certain orbits; these orbits Bohr called 'stationary states', because only in these orbits can the electron remain an appreciable time.*

When the electron leaves one stationary state it must pass instantaneously to another; and if these two states have energies E' and E'', so that $E' > E''$, then Bohr's second postulate adds that a quantum of monochromatic light will be emitted, the frequency ν' being determined by the equation

$$h\nu' = E' - E'', \qquad \text{......(2·3)}$$

where h is Planck's quantum of action $6{\cdot}55 . 10^{-27}$ erg sec. Sub-

* Bohr, *PM*, 1913, **26** 1, 476, 857.

stituting the wave-number ν for the true frequency ν', the equation reduces to

$$\nu = \frac{E'}{hc} - \frac{E''}{hc}, \qquad \text{......(2·4)}$$

where c is the velocity of light. As in the Balmer formula the wave-number of the emitted line is here expressed as the difference of two terms, but the two may be made identical by equating E'/hc to R/n_1^2 or $-R/n_2^2$. As the strongest hydrogen lines arise between states having small values of n, these are presumably the most stable and have the lowest energy values; accordingly, the two expressions are best made equivalent by writing the energy E_l of a stationary state as

$$E_l = -\frac{hcR}{n^2}. \qquad \text{......(2·5)}$$

This choice will make the energy zero when n is infinite, that is presumably when the electron has been removed to an infinite distance and the atom has been ionised; making this explicit we write E_l instead of E, since E is now measured from the limit.

Continuing on the assumption that an electron travelling in a permitted orbit obeys the classical laws, Bohr developed a mechanical model, which is so successful in co-ordinating the more important facts that it is still worth study to-day, even though Heisenberg has given a more general treatment.

To make the argument as simple as possible, suppose that the mass of the nucleus is infinitely great compared with that of the electron, and that the velocity of the latter is very small compared with that of light. If the accelerated electron does not emit radiation, it will describe an ellipse with the nucleus in one focus as Kepler's first law requires. The frequency of revolution ω and the major axis of the ellipse, $2a$, are given by the equations

$$\omega^2 = \frac{2W^3}{\pi^2 e^4 m_e}, \quad 2a = \frac{e^2}{W}, \qquad \text{......(2·6)}$$

where e and m_e are the charge and mass of the electron, while W is the work which must be added to the system in order to remove the electron to an infinite distance. The equations (2·6) show that both the frequency of revolution and the magnitude of the major axis depend only on the energy. Assuming as in equation (2·5) that the energy is zero when the atom is ionised,

$$W = -E_l = \frac{hcR}{n^2}. \qquad \text{.....(2·7)}$$

Combining equations (2·5) and (2·6) the frequency of revolution ω_n of the electron in the nth orbit appears as

$$\omega_n{}^2 = \frac{2}{\pi^2}\frac{R^3h^3c^3}{e^4m_en^6}. \qquad \ldots\ldots(2{\cdot}8)$$

A particular line of the spectrum arises from a transition between two stationary states in each of which the electron revolves with a certain frequency; but no simple relation between these frequencies of revolution and the frequency of the emitted radiation is to be expected. Long before Bohr wrote, however, Planck had shown that the quantum theory of black body radiation agrees closely with the classical theory, when the electromagnetic oscillations are sufficiently slow; so Bohr plausibly assumed that this correspondence extends to the mechanics of the hydrogen atom. If it does, a transition between the nth and $(n+1)$th states should give rise to radiation of the frequency predicted by the classical theory when n is sufficiently large; the frequency of the radiation will then be simply ω.

According to the quantum theory the frequency of the radiation arising from the jump between the $(n+1)$th and nth stationary states is given by

$$\nu' = Rc\left\{\frac{1}{n^2} - \frac{1}{(n+1)^2}\right\}.$$

When n is very large, this expression is approximately equal to

$$\nu' = \frac{2Rc}{n^3}.$$

But according to the classical electrodynamics this is the frequency of revolution ω_n.

This value of ω_n will satisfy equation (2·8) if, and only if,

$$R = \frac{2\pi^2e^4m_e}{ch^3}. \qquad \ldots\ldots(2{\cdot}9)$$

Thus Bohr by using here the quantum theory and there the classical theory mixed a hotch-potch which, when properly baked, formed itself into an expression relating R to the other physical constants. When the known values of e, m_e and h were introduced into this expression, the value obtained was 109700 cm.$^{-1}$; and this agreed very well with the spectroscopic value of 109678 cm.$^{-1}$ In early years this numerical agreement was the most powerful argument in favour of the theory.

The above argument follows closely that of Bohr's original paper. It is not so logical an argument as could be constructed to-day, but it keeps the physical meaning of each step clear, and offers a very elementary example of the correspondence between wave theory and quantum theory. This 'correspondence' has been developed into a principle by Bohr, and has been used with great success in deducing selection rules, intensity relations and the like.*

Further confirmation of Bohr's theory of the hydrogen atom may be obtained by examining the energy and size of the atom predicted. The atom will be most stable when W assumes its greatest value; this is $W = chR$, corresponding to $n = 1$. Under normal conditions most atoms will probably be in this state, and only a negligible number in the state having $n = 2$. To this then corresponds the observation that lines of the Balmer series do not appear in absorption spectra in the laboratory, though they appear reversed in the spectra of certain stars.

Again the major axis of the ellipse in which the electron moves is

$$2a = \frac{e^2}{chR} n^2 = 1{\cdot}1 \, . \, 10^{-8} n^2 \text{ cm.} \qquad \text{......(2·10)}$$

When n is small the values obtained from this equation agree well with those obtained from the kinetic theory of gases, but atoms in the higher states are to be expected only at low pressures. When $n = 20$, the diameter of the atom should be $4{\cdot}4 \, . \, 10^{-6}$ cm. according to Bohr, and atoms will be at this mean distance apart only when the pressure is reduced to 0·3 mm. of mercury, so that transitions from this state to another can hardly be expected to give rise to lines of measurable intensity until the pressure has been reduced beyond this.

In the laboratory only the first four lines of the Balmer series are usually observed, but Whiddington† has shown that when the pressure is reduced from 1 mm. to 10^{-3} mm. of mercury, the number of lines can easily be raised from four to twenty. Further, there seems to be some evidence that the lower the pressure in a star the greater the number of Balmer lines observed. In ζ Puppis Pickering actually observed 33 lines.

* A good account of the correspondence principle is found in Andrade, *The Structure of the Atom*, 1927, 206.

† Whiddington, *PM*, 1923, **46** 605.

4. Ionised helium

Bohr's simple theory should be applicable to ionised helium, which like hydrogen consists of a single electron rotating about a positively charged nucleus. If the theory be revised for a nuclear charge of Ze, it indicates that the spectrum will be represented by

$$\nu = \frac{2\pi^2 Z^2 e^4 m_e}{ch^3}\left\{\frac{1}{n_1^2} - \frac{1}{n_2^2}\right\} = Z^2 R_{\mathrm{H}}\left\{\frac{1}{n_1^2} - \frac{1}{n_2^2}\right\}, \quad \ldots\ldots(2\cdot 11)$$

where R_{H} is identical with the quantity previously written R.

Pickering* long ago discovered in ζ Puppis certain lines which can be nearly, but not precisely, represented by this formula when $Z=2$ and $n_1=4$; while later Fowler,† passing a condensed discharge through helium, observed a second series having $n_1=3$. At first both series were attributed to hydrogen, and only when Bohr‡ had shown how the lines might be explained as part of the spark spectrum of helium was this view confirmed in the laboratory.§

Though alternate lines of the Pickering series lie close to lines of the Balmer series, they do not coincide; agreement with equation (2·11) can however be improved, if a new constant R_{He} is substituted for the old one R_{H}. This change Bohr ascribed to the different masses of the two nuclei. The above discussion has assumed that the mass of the nucleus is infinite; if in fact the mass is finite and equal to M, the electron will revolve not about the nucleus, but about the centre of gravity of nucleus and electron. Allowance can be made for this by writing in all equations instead of the mass m_e the quantity $\frac{M}{M+m_e}m_e$. Accordingly

$$\frac{R_{\mathrm{He}}}{R_{\mathrm{H}}} = \frac{1+m_e/M_{\mathrm{H}}}{1+m_e/M_{\mathrm{He}}}. \quad \ldots\ldots(2\cdot 12)$$

The value of R_{He} is $109722\cdot 40 \pm 0\cdot 05$ cm.$^{-1}$, which gives the ratio M_{H}/m_e the value 1843. This is in satisfactory agreement with the deflection of an electron in a magnetic field, for

* Pickering, *AJ*, 1896, **4** 369; 1897, **5** 92.

† Fowler, A., *Roy. Astron. Soc.*, *M.N.* 1912, **73**, 62.

‡ Bohr, *PM*, 1913, **26** 1.

§ Evans, *PM*, 1915, **29** 284.

the latter shows the hydrogen atom 1838 times as heavy as an electron.*

Series homologous with those observed in hydrogen and ionised helium are to be expected in any ion, which has only one electron, and in fact five lines homologous with the Lyman series have been observed in the second spark spectrum of lithium, sometimes written Li III.†

5. The wave mechanics

Although the wave mechanics was born of abstruse mathematics, some of the ideas may be expressed in quite simple terms.‡

Ten years ago physicists were impaled on a dilemma; the wave theory and only the wave theory would explain interference and diffraction; the quantum theory and only the quantum theory would account for interchange of energy between radiation and matter. Moreover, the quantum theory of the atom which gradually developed on Bohr's foundation contained so many *ad hoc* assumptions, that only those who did not fully understand it could feel satisfied. In 1925, however, the genius of De Broglie§ showed that these two antagonistic theories might be woven into one whole, and in doing so rendered unnecessary the arbitrary assumptions of the old quantum theory. Most physicists would have been content had De Broglie shown that the two views were compatible when light alone is considered, but Compton had already shown that when X-rays scatter electrons, the quanta or photons behave as if they are solid particles having a definite momentum; and so De Broglie cast his net far wider and boldly asserted that since matter can carry momentum, matter like light will behave under suitable conditions as if it is waves.

* Comparison of the position of H_α in hydrogen and heavy hydrogen offers an even more accurate method of measuring M_H/m_e. Shane and Spedding, *PR*, 1935, **47** 33.

† Gale and Hoag, *PR*, 1931, **37** 1703*a*.

‡ This section is based largely on West, *Science Progress*, 1931, **25** 622.

§ Professor R. H. Fowler has kindly pointed out that de Broglie's ideas were implicit in Heisenberg's matrix mechanics. History relates that Schrödinger showed the matrix mechanics capable of a wave interpretation, thus simplifying it, and that Dirac generalised the result. De Broglie's wave theory is important to-day, but it exerted little influence on the course of development at the time.

Consider how these ideas may be applied to the hydrogen atom. If the revolving electron behaves as a wave of length λ, the circumference of the orbit must be an exact multiple of the wave-length, or the wave will interfere destructively with itself. The condition therefore that a stationary wave system shall be formed appears to be

$$2\pi a = n\lambda, \qquad \ldots\ldots(2{\cdot}13)$$

where n is integral, and the orbit is supposed circular.

In order to use this relation, however, the wave-length of the electron must be discovered. To obtain this, consider again Compton's discovery that when a photon strikes an electron, momentum is conserved, the momentum of an electron being $m_e v$, and of a photon $h\nu/c$.

The reasons for taking $h\nu/c$ as the momentum of a photon are various; thus Einstein's work on the special theory of relativity led him to state that any quantity of energy E has associated with it a mass E/c^2; but the energy of a photon is $h\nu$ and it travels with velocity c, so that it should carry momentum $\frac{h\nu}{c^2}.c = \frac{h\nu}{c}$. And this links up satisfactorily with Maxwell's statement that the pressure exerted by isotropic radiation is $E/3$ where E is now the energy of unit volume; for if there are n photons per unit volume, and these are considered as divided into three equal groups travelling parallel to the three faces of a box, the number striking unit area on one face is $nc/3$ per second; accordingly the momentum destroyed per second is $\frac{h\nu}{c}.\frac{nc}{3} = \frac{1}{3}nh\nu$, which is $E/3$, as Maxwell's theory and Stefan's experimental law both require.

If the momentum of a photon is $h\nu/c$, however, it may equally well be written h/λ; and if this relation is valid also for an electron then

$$m_e v = \frac{h}{\lambda},$$

where v is the velocity with which the electron is travelling.

The wave-length of an electron is thus $h/m_e v$, and the condition for a stationary wave system simplifies to

$$2\pi a = \frac{nh}{m_e v}.$$

This will determine the radii a of the permitted orbits, if a further relation is obtained between a and v, as it may be by following Bohr's derivation and applying the classical laws to the motion of the electron in a circular orbit; for these laws show that if the nuclear charge is Ze, then

$$\frac{Ze^2}{a^2}=\frac{m_e v^2}{a} \quad \text{or} \quad v^2 a=\frac{Ze^2}{m_e}. \qquad \text{......(2·14)}$$

Thus the condition for a stationary wave system becomes

$$a=\left\{\frac{nh}{2\pi m_e}\right\}^2 \Big/ \frac{Ze^2}{m_e} \quad \text{or} \quad 2a=\frac{n^2h^2}{2\pi^2 m_e Ze^2}, \qquad \text{......(2·15)}$$

and this is the expression obtained much earlier by Bohr.

BIBLIOGRAPHY

Grotrian. *Graphische Darstellung der Spektren.* 1928.

CHAPTER III

THE ALKALI DOUBLETS

1. Detection of series

In the elucidation of simple spectra early workers depended almost entirely on the discovery of series of lines. These series have the same general appearance as the series of hydrogen; the lines decrease in intensity and lie nearer and nearer together as the eye moves from the red to the violet. But this definition alone is seldom sufficient to identify the lines of a series, for in all spectra, except hydrogen, different series overlap one another. This is illustrated in Fig. 3·1, which shows the spectrum of lithium.

Sometimes the appearance of the lines helped early workers, and so the terms 'sharp' and 'diffuse' were used to designate particular series in the alkalis. Again in lithium, which we may select as a type for this description, the lines of a third series are particularly intense and are easily absorbed by cold lithium vapour, so that it is known as the 'principal' series. Often, too, lines of this series appear 'self-reversed'; that is to say the normal line has a narrow dark band down the centre of it, an appearance which arises when the centre of the broad line emitted by the hot core of the discharge is re-absorbed by the cooler layers outside; the line emitted will naturally be broader than that absorbed, because the Doppler effect of a moving source will broaden a line more the greater the velocity of the source, and therefore the higher the temperature.

Any series of hydrogen may be written in the form

$$\nu = T_l - \frac{R}{n^2}, \qquad \text{......(3·1)}$$

where T_l is the wave-number of the limit. Rydberg showed that this may be generalised for series of other elements in the form

$$\nu = T_l - \frac{R}{(m+a)^2}, \qquad \text{......(3·2)}$$

where a is a constant characteristic of the series and m assumes a series of integral values as n did. m is commonly called the 'serial' number. Rydberg's constant R has the same value in all

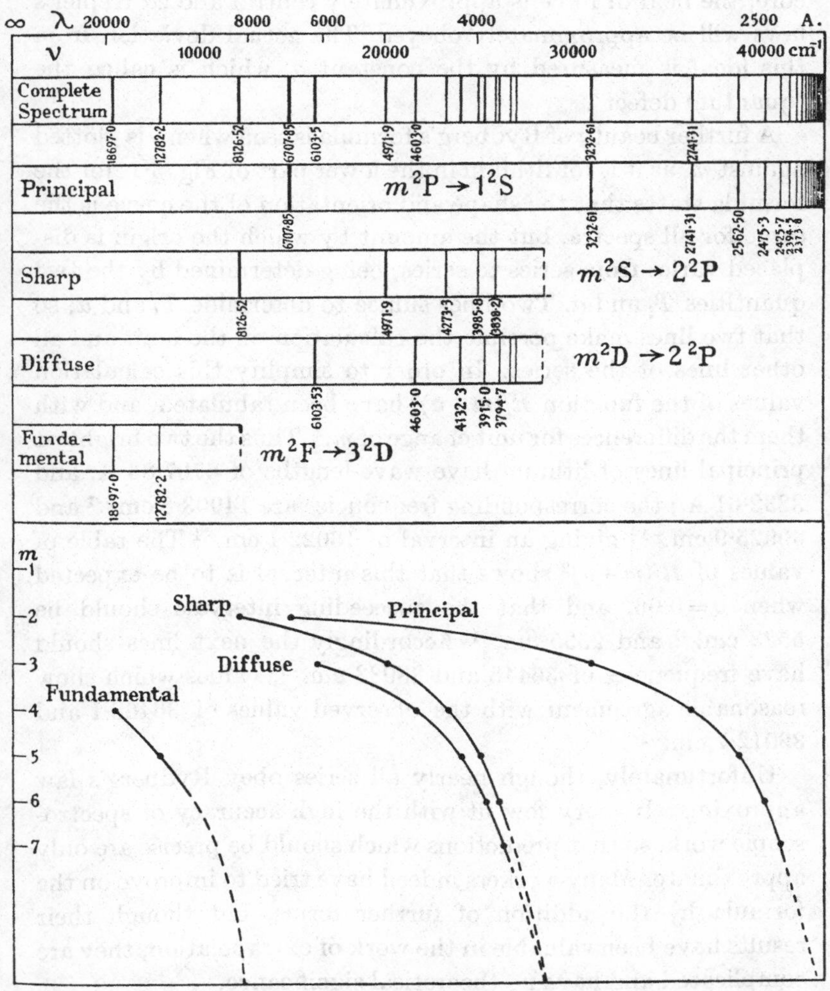

Fig. 3·1 Series in lithium. The lower part of the figure illustrates Rydberg's series formula.

spectra, if small deviations in a few light atoms are for the present ignored. In terms of the model of the atom, this seems to mean that the greater part of the orbit of the series electron lies outside the orbits of the remaining electrons, the latter with the nucleus being commonly known as the 'core'. In the region outside the core, the field of force is approximately central and so Kepler's laws will be approximately obeyed. The actual deviation from this ideal is measured by the constant a, which is called the 'quantum defect'.

A further beauty of Rydberg's formula is seen when ν is plotted against m, as it is for lithium in the lower part of Fig. 3·1; for the formula states that the shape and orientation of the curve is the same for all spectra, but the amount by which the origin is displaced varies from series to series, being determined by the two quantities T_l and a. Two lines suffice to determine T_l and a; so that two lines make possible the calculation of the limit and all other lines of the series. In order to simplify this calculation values of the function $R/(m+a)^2$ have been tabulated, and with them the differences for unit change of m.* Thus the two brightest principal lines of lithium have wave-lengths of 6707·85 A. and 3232·61 A.; the corresponding frequencies are 14903·8 cm.$^{-1}$ and 30925·9 cm.$^{-1}$† giving an interval of 16022·1 cm.$^{-1}$ The table of values of $R/(m+a)^2$ shows that this interval is to be expected when $a=0{\cdot}96$, and that the succeeding intervals should be 5522 cm.$^{-1}$ and 2535 cm.$^{-1}$ Accordingly the next lines should have frequencies of 36448 and 38983 cm.$^{-1}$, values which show reasonable agreement with the observed values of 36468·1 and 39012·7 cm.$^{-1}$

Unfortunately, though nearly all series obey Rydberg's law approximately, very few fit with the high accuracy of spectroscopic work, so that predictions which should be precise are only approximate. Many workers indeed have tried to improve on the formula by the addition of further terms, but though their results have been valuable in the work of extrapolation, they are complicated and have no theoretical significance.

Accordingly, the modern spectroscopist seeks values of T_l and

* See Appendix.

† The value of ν is not precisely $10^8/\lambda$, because I follow Fowler in giving the wave-length in air, and the wave-number *in vacuo*. The corrections to be applied are tabulated in his *Report on Series in Line Spectra*, 1922, 2 and 81.

PLATE I

1*a*. Balmer series in hydrogen. As one passes from red to violet, the intensities decrease and the lines lie closer together.

1*b*. Three doublets of sodium. The difference which caused one series to be named 'sharp' and another 'diffuse' is here apparent; the lines at 6161 A. are finer than those at 5688 A.; the self-reversal of the principal doublet at 5890 A. is also characteristic. The 5890 A. doublet is so easily excited that it appears in the iron arc, though sodium is there present only as an impurity.

1*c*. Doublet intervals of the alkalis. Comparison of the first diffuse doublet of sodium with the first principal doublets of potassium and rubidium shows how the interval increases with the atomic number. (Photograph by Col. E. H. Grove-Hills.)

2. Series in lithium. In all series the lines lie nearer together and grow fainter as one passes from red to violet.

3. Series in sodium. The doublets are not resolved.

4. Triplet series in zinc. The fine lines of the sharp series contrast with the blurred lines of the diffuse; the lines decrease in intensity and converge to a limit as one passes into the ultra-violet. The principal series lies in the red.

5. Triplet series in cadmium. The triplets are wider here than in zinc, so that they more frequently overlap. Zinc appears as an impurity in the cadmium spectrum, and cadmium as an impurity in zinc.

All the above photographs were lent by Prof. A. Fowler. They have been previously reproduced in his *Series in Line Spectra*.

Plate I

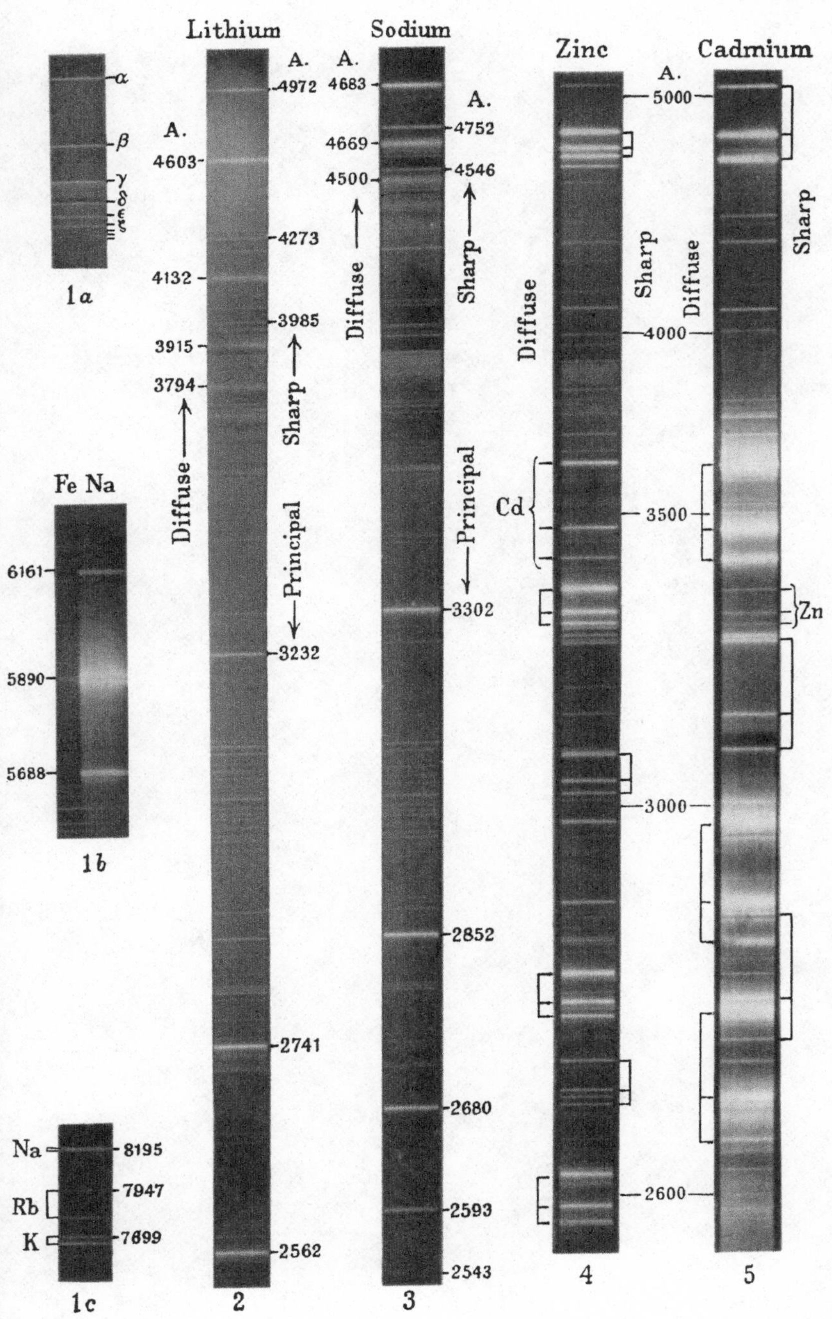

α
β
γ
δ
ε
ζ
1a
Fe Na
6161
5890
5688
1b
Na
8195
Rb
7947
K
7699
1c
Lithium
A.
4972
4603
4273
4132
3985
3915
3794
Diffuse
Sharp
Principal
3232
2741
2562
2
Sodium
A.
4683
4752
4669
4500
4546
Diffuse
Sharp
Principal
3302
2852
2680
2593
2543
3
Zinc
A.
5000
Diffuse
Sharp
4000
Cd
3500
3000
2600
4
Cadmium
Diffuse
Sharp
Zn
5

a which render the deviations between Rydberg's equation and the observed wave-lengths as small as possible.* Knowing that all the lines of a series behave similarly in a magnetic field, he uses this as his touchstone to decide whether a given line belongs to a certain series. Some series law is essential indeed if the limit and the ionisation potential are to be estimated, but the Rydberg formula helped out by empirical rules is accurate enough for this.

2. Rydberg–Ritz combination principle

The wave-number of a line of the sharp series may be written

$$\nu = T_l - \frac{R}{(m+\mathrm{s})^2} \quad \text{or} \quad \nu = \mathrm{S}_l - m\mathrm{S}. \qquad \text{......(3·3)}$$

Both these expressions show the wave-number as the difference of two terms, but they employ different notations. The equation on the left is the simple Rydberg formula, in which s has been substituted for a to show that the variable term belongs to the sharp series.

The fact that the lines of a series fit a Rydberg formula closely is of great value in analysis, but in theory the sequence to which a term belongs and its energy value are both more important than this closeness of fit. Accordingly, a notation which emphasises the sequence and ignores the series formula, as that on the right does, is to be preferred. Moreover in the equation

$$\nu = \mathrm{S}_l - m\mathrm{S}$$

the terms S_l and $m\mathrm{S}$ may be taken as the energies† of the stationary states in which the electron ends and begins its jump. Since these energies are measured down from the limit of the series, that is from a hypothetical term which should be written $\infty\mathrm{S}$, the positive term values represent negative energies, and numerically $\mathrm{S}_l > m\mathrm{S}$. In the alkalis the brightest line of the sharp series is $\mathrm{S}_l - 2\mathrm{S}$, and the next brightest $\mathrm{S}_l - 3\mathrm{S}$. The greater m, the greater the energy of the atom but the smaller the term, so that the line $\mathrm{S}_l - 2\mathrm{S}$ is of lower frequency than $\mathrm{S}_l - 3\mathrm{S}$.

* For some empirical rules, Russell, *AJ*, 1927, **66** 233. Fowler, A., *Report on Series in Line Spectra*, 1922, 31 f.

† The purist will rightly object that energy cannot be measured in cm.$^{-1}$; the terms should be multiplied by hc.

Exactly analogous to the sharp series are the lines of the principal, diffuse and fundamental series, which are written

$$\nu = \mathrm{P}_l - m\mathrm{P}, \quad m \geqslant 2,$$
$$\nu = \mathrm{D}_l - m\mathrm{D}, \quad m \geqslant 3,$$
$$\nu = \mathrm{F}_l - m\mathrm{F}, \quad m \geqslant 4.$$

The name 'fundamental' was given to the last of these series because the terms have a very small quantum defect; early workers took this as showing a resemblance to the simple or 'fundamental' spectrum of hydrogen; unfortunately to-day the fact receives quite a different explanation and the series can be regarded as fundamental in no physical sense; on the continent the series is named after Bergmann, but as the symbol F is everywhere accepted, a change of nomenclature seems hardly justified.

In the pragmatic spirit which seeks to eliminate all useless theory, the spectroscopist commonly uses the effective quantum number $n^\star$, defined by the relation

$$\mathrm{T} = R/n^{\star 2}, \qquad \text{......(3·4)}$$

where T is any term such as $m\mathrm{P}$ or P_l. This makes $n^\star$ equal to the $(m+a)$ of the Rydberg equation, and the notation is useful in that m and a cannot be separated by experiment. Thus in Fig. 3·2

m	Sharp	Principal	Diffuse	Fundamental
1	1·59			
2	2·60	1·96		
3	3·60	2·96	3·00	
4	4·60	3·95	4·00	4·00
5	5·60	4·95	5·00	5·00
6	6·60	5·95	6·00	6·00

m	Sharp	Principal		Diffuse	Fundamental
		$J=\frac{1}{2}$	$J=1\frac{1}{2}$		
1	1·87				
2	2·92	2·33	2·36		
3	3·93	3·37	3·41	2·55	
4	4·94	4·39	4·42	3·54	3·98
5	5·94	5·40	5·43	4·53	4·98

Fig. 3·2. Effective quantum numbers of lithium (above) and caesium (below).

are given the values of the effective quantum number of the terms of the four lithium series, and on the left are set the values of m allotted by Paschen and Götze.

The allotment of the serial number caused considerable worry to early workers. It may be chosen so that a is always numerically less than half, or so that a is always negative, or again so that the lowest terms of the four sequences are 1S, 2P, 3D and 4F. Each has its peculiar advantage, but the third, favoured by Paschen and Götze, will be followed here, though even it will be subject to certain exceptions to be pointed out from time to time. The Paschen and Götze notation at least marks out the lowest terms, and does not usually involve large values of the quantum defect; in lithium, for example, a assumes the values of 0·60, −0·04, 0 and 0 in the four term sequences. The values of m will be seen later to vary uniformly with the values of the chief quantum number n, the two differing by a constant value in any one sequence.

A further simplification is still possible. The limit of the sharp series is identical with the lowest term of the P sequence which we have agreed to write 2P; that is $S_l = 2P$. And similarly experiment shows that $D_l = 2P$ and $F_l = 3D$, while P_l coincides with a term extrapolated from the sharp sequence and written accordingly 1S. Thus the series may be revised to read:

Principal series:	$1S - mP$	or	$mP \rightarrow 1S$,	$m = 2, 3, \ldots,$
Sharp series:	$2P - mS$	or	$mS \rightarrow 2P$,	$m = 2, 3, \ldots,$
Diffuse series:	$2P - mD$	or	$mD \rightarrow 2P$,	$m = 3, 4, \ldots,$
Fundamental series:	$3D - mF$	or	$mF \rightarrow 3D$,	$m = 4, 5, \ldots.$

These are very important, for the formulae hold not only in lithium and the other alkalis, but in all simple spectra. The first line of the sharp series might be 2P–1S, but in the alkalis the 1S term is always greater than the 2P; or in words, the atom is more stable in the 1S than in the 2P state. Of the alternative notations, $1S - mP$ is part of an equation from which the left-hand side ν has been omitted; while in $mP \rightarrow 1S$ the higher level is placed on the left and the lower on the right, the arrow showing the direction of the transition; in absorption the same line would be written $mP \leftarrow 1S$.

In all simple spectra terms are customarily measured from the limit of a sequence; this is reasonable when the ground term lies so deep that all its combinations lie in the far ultra-violet; but elsewhere the practice is to be deplored, because the series limit obtained by the extrapolation of a sequence is necessarily less accurate than the term differences obtained direct from the wave-lengths. In recent analyses of complex spectra the terms are usually measured up from the ground term, and when this is done the $m\mathrm{P} \rightarrow 1\mathrm{S}$ notation is preferable to $1\mathrm{S}$–$m\mathrm{P}$. As moreover the arrow notation is now standard in molecular spectra, it will be adopted hereafter, and term symbols will be taken to denote energies measured up from some arbitrary zero.

The identity of the series limits with certain terms introduces the first evidence of the 'combination principle', which states that the wave-number ν of any line may be written as the difference of two terms,

$$\nu = \mathrm{T}'' - \mathrm{T}'.$$

For this principle developed by Rydberg and Ritz* is of value only if the number of terms postulated is less than the number of lines observed.

All the evidence goes to show that the combination principle is precise, as indeed it must be if energy is to be conserved, and so in compiling lists of standard wave-lengths in a region difficult to photograph, such as the infra-red, the value calculated from the difference of two visible lines is often preferred to direct measurement.

The series of lithium may be well represented by the scheme of atomic levels shown in Fig. 3·3. The energy levels might be represented by horizontal lines stretching right across the page, but to simplify the diagram the levels belonging to different sequences are arranged on different verticals. The broken diagonals represent the combinations which have been observed, and in the break is inserted the wave-length of the resulting line. The thickness of the diagonal is roughly proportional to the intensity of the line.

All but one of the lines shown in Fig. 3·3 belong to the four series

* Rydberg, *AJ*, 1896, **4** 91; Ritz, *PZ*, 1908, **9** 521; Ritz, *AJ*, 1908, **28** 237.

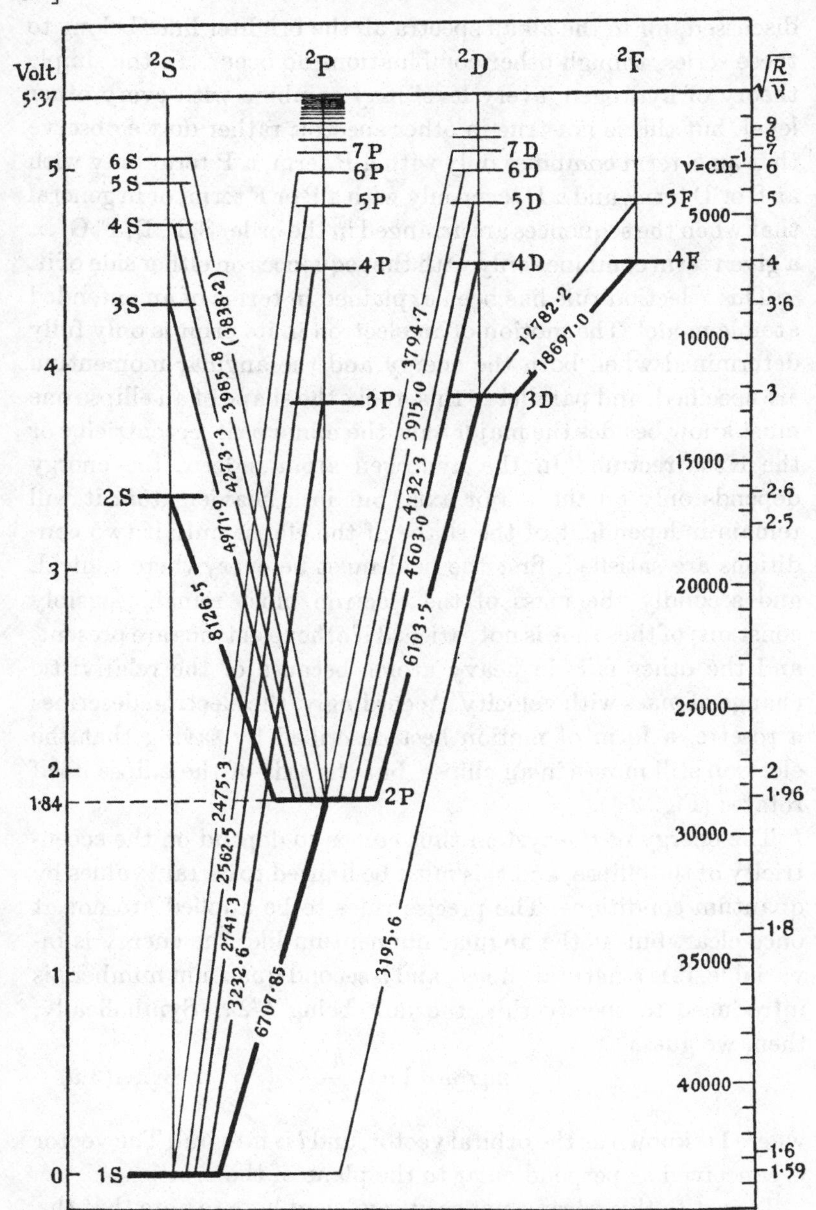

Fig. 3·3. Level diagram of lithium. After Grotrian, *Graphische Darstellung der Spektren.*

discussed, for in the alkali spectra all the brighter lines belong to these series, though other combinations do occur. In the simple theory of hydrogen, every level may combine with every other level, but this is not true in other spectra; rather do we observe that an S term combines only with a P term, a P term only with an S or D term, and a D term only with a P or F term; or in general that when the sequences are arranged in the order S, P, D, F, G, ... a given term combines only with the sequences on either side of it.

This selection rule has been explained in terms of an extended atomic model. The motion of an electron in its orbit is only fully determined when both the energy and the angular momentum are specified, and parallel to this, to fix the shape of an ellipse one must know besides the major axis, the minor axis, eccentricity or the latus rectum. In the hydrogen atom indeed, the energy depends only on the major axis, but in a heavier atom it will remain independent of the shape of the ellipse only if two conditions are satisfied; first the field must be everywhere central, and secondly the mass of the electron must remain sensibly constant; of these one is not satisfied if other electrons are present, and the other fails in heavy atoms because of the relativistic change of mass with velocity. Accordingly, the electron describes a rosette, a form of motion best described by saying that the electron still moves in an ellipse, but the axis of the ellipse itself rotates (Fig. 3·4).

The energy of the system thus comes to depend on the eccentricity of the ellipse, and this must be limited to certain values by quantum conditions. The precise rules to be applied are not at once clear, but as the angular momentum like the energy is invariable, it is a natural choice, and a second quantum number l is introduced to specify this, the unit being $h/2\pi$. Symbolically, then, we guess

$$m_e r^2 \omega = \mathbf{l} = l \cdot \frac{h}{2\pi}, \qquad \text{......(3·5)}$$

where $\mathbf{l}$ is known as the orbital vector, and l is integral. The vector is conceived as perpendicular to the plane of the orbit.

To justify this guess is not easy; sufficient here to note that the quantum mechanics leads to the same result, and to point out one or two simple consequences. Using the correspondence principle, Bohr showed that if the model is correct, then in any jump l can change only by unity; that is $\Delta l = \pm 1$. Clearly the four simple series are in accord with this selection rule, if terms of the S sequence have $l = 0$, and the P, D and F terms have l values of 1, 2 and 3, respectively. This selection rule allows however other lines, and in fact series such as the mS → 3P and mD → 3P found in lithium do conform. Occasionally it is true lines are found

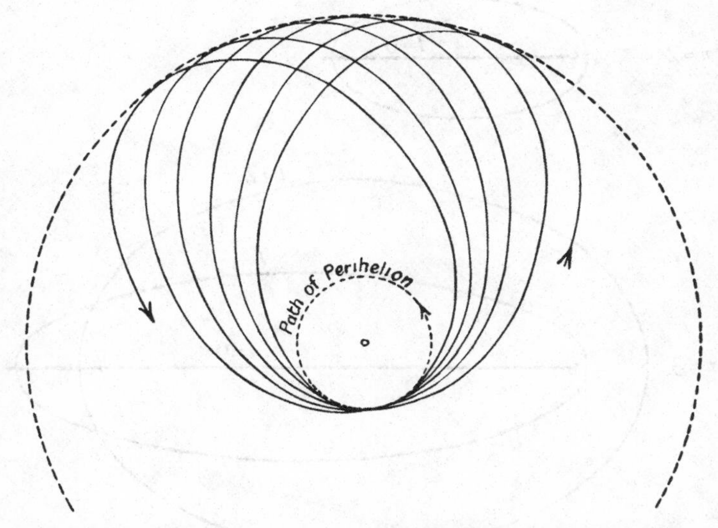

Fig. 3·4. Orbit of an electron, which penetrates within the core.

which violate the selection rule, the 3P → 2P line of lithium for example; but such lines tend to occur only in an electric field and even then they are weak. A more interesting exception is the mD → 1S series, which has been observed when the arc is first struck and before the normal spectrum has had time to develop;* modern theory regards this series as an example of quadripole radiation, and links it with certain forbidden nebular lines.

The model has the further merit of co-ordinating some other observations. As the major axis of an elliptical orbit depends only

* Shrum, Carter and Fowler, H. W., *PM*, 1927, **3** 27; Grotrian, *Graphische Darstellung der Spektren*, 1928, **1** 54, and Chapter XXI of this book.

on the energy, so a simple analysis shows that the latus rectum depends only on the angular momentum; indeed the expressions for the two lengths are similar, for as the major axis is $\frac{Ze^2}{chR}n^2$,

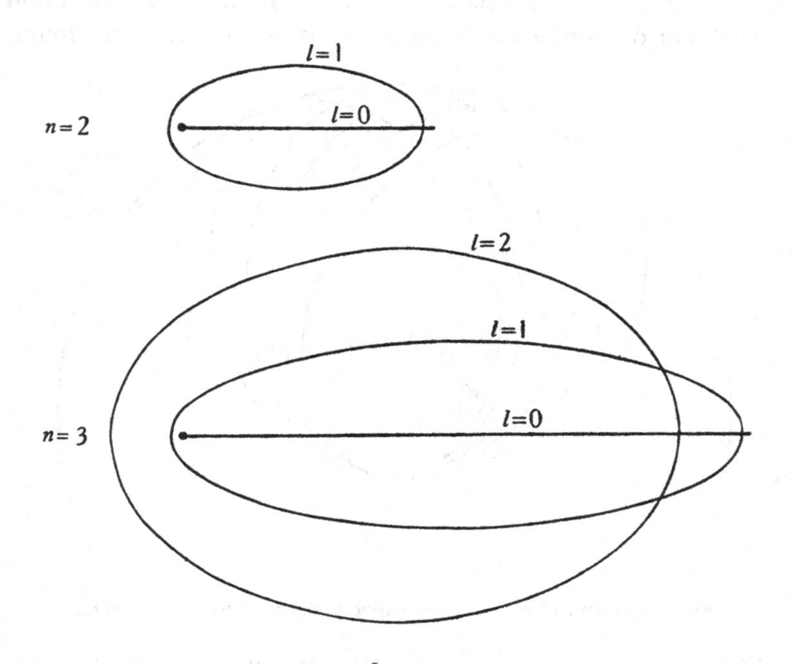

Scale —— 1·06 10^{-8} cm.

Fig. 3·5. Electron orbits of a hydrogen atom drawn to scale.

so the latus rectum is $\frac{Ze^2}{chR}l^2$. Moreover, the minor axis, $2b$, of an ellipse is the geometric mean of the major axis, $2a$, and the latus rectum, so that $\frac{b}{a}=\frac{l}{n}$. The orbits permitted by the two quantum conditions are drawn to scale in Fig. 3·5; this shows that of orbits having the same value of n, those pass nearer to the nucleus which have the smaller orbital vector. Now the deviation from simple theory will presumably be greater when the valency electron penetrates more deeply into the core; it will be greater in an S orbit than in a P.* This deviation is measured roughly by the quantum defect, a, which in lithium, as in most other spectra, decreases as l increases. In fact the quantum defect is almost zero in the D and F terms, so that these orbits must approximate to those of hydrogen. Again, the deviation and the quantum defect will presumably increase as the atomic number and the size of the core increase. And in agreement Fig. 3·2 shows that the quantum defects of caesium are much larger than those of lithium.

The very simplicity of the vector model is repugnant to the principles of the wave mechanics, for the position of an electron at any time and its velocity cannot both be accurately determined; but this model does in fact serve to co-ordinate many experimental facts, and provides in addition a simple picture of what is happening, a picture which can hardly fail to help those who cannot face the mathematical analysis. And if we find it hard to picture the electron passing through the nucleus as it does in the model when l is 0, we can at least remember that we expect two wave-trains to cross.†

3. Doublet series

The spectroscopic series have been introduced by considering a singlet spectrum; but as the true singlet system of the alkaline earths occurs in conjunction with a triplet system, the doublet system of lithium has been used with sufficiently low resolution. Theory leaves no doubt that the lines of lithium are doublets like those of the other alkalis, but the separation is so small that experiment has failed to resolve more than a few of them. The separation of two doublet levels varies roughly as the square of the atomic number, so that a separation which is small in sodium will be much larger in caesium. The facts, which are summarised below, are beautifully explained by the assumption that the P, D

* For an admirable discussion of 'penetrating' and 'non-penetrating' orbits, see Ruark and Urey, *Atoms, molecules and quanta*, 1930, 197.

† Darwin, C. G., in *New conceptions of matter*, 1931, gives a delightfully lucid discussion.

and F levels are double while the S level in the alkalis, as in all spectra, remains single.

Since some atoms, though not the alkalis, are capable of emit-

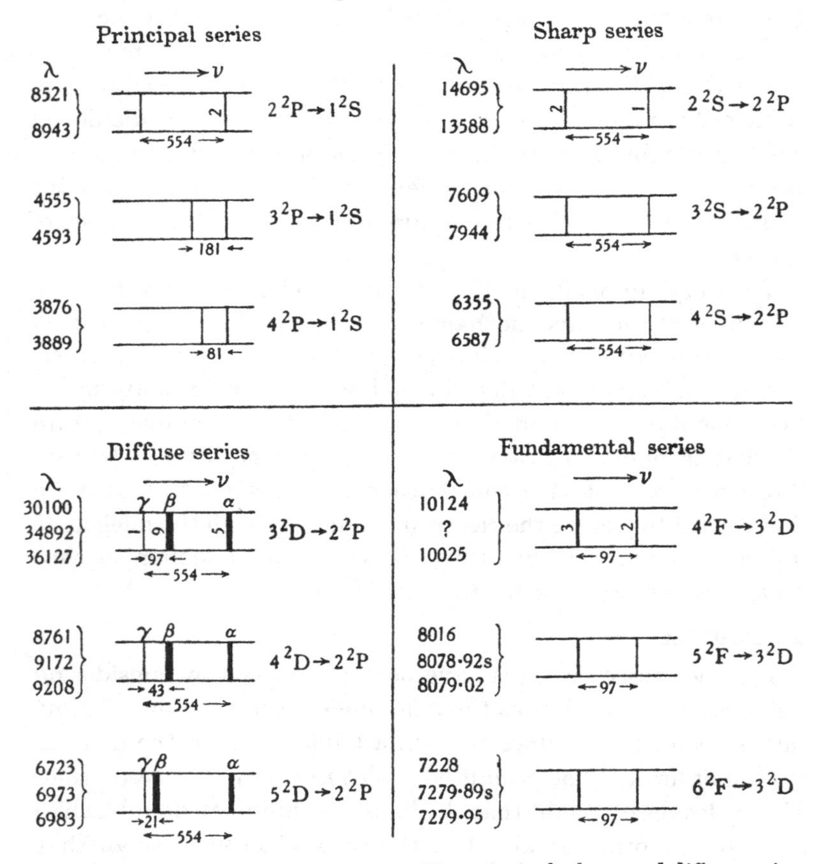

Fig. 3·6. Doublet structure of caesium. The principal, sharp and diffuse series are all drawn with the same frequency scale, the intervals being shown below in cm.$^{-1}$ The relative intensities of the components of each doublet are shown roughly by the widths of the lines, and more accurately by numbers set beside the first doublet of each series.

ting two systems of lines having different multiplicities, it is convenient to denote the system to which a term belongs by a small prefix placed at the top of the term symbol; thus the doublet, S, P and D terms are written ^2S, ^2P and ^2D. Further, to distinguish the two P, D and F levels, we add a suffix and write $^2P_{\frac{1}{2}}$ and $^2P_{1\frac{1}{2}}$, $^2D_{1\frac{1}{2}}$ and $^2D_{2\frac{1}{2}}$, $^2F_{2\frac{1}{2}}$ and $^2F_{3\frac{1}{2}}$. The reason why the suffixes are not always written 1 and 2 will appear later; for the present the notation put forward may be regarded as a harmless idiosyncrasy of the author.

The actual appearance of the doublets in the four series, summarised by Fowler in his *Report on Line Spectra*, are set out below and are illustrated in Fig. 3·6.

A. In the SHARP series the less refrangible components of the pairs are the stronger and the separation of the components, when expressed in wave-numbers, is constant. Reverting to Rydberg's formula, these series must have the same value of a but different values of T_l; the separation of the limits will be the separation of each doublet.

B. In the PRINCIPAL series the more refrangible components of the pairs are the stronger. The first pair has the same separation as that of the sharp series, but the components approach each other as the serial number increases and the two series have the same limit. Reverting again to Rydberg's formula, these two series will have the same value of T_l, but different values of the quantum defect a.

C. The DIFFUSE series was at first thought to be governed by the rules already given for the sharp series, the separation of the doublets being the same in both series; but more accurate work has shown that the less refrangible line of the pair has a faint companion or 'satellite'. While the brighter component is displaced from its previously accepted position, the satellite obeys the law of constant separation. The chief line lies on the more refrangible side of the satellite and approaches it as the serial number increases, the common limit being identical with that of the less refrangible line of the sharp series.

In days gone by the satellite was thought to be a blot on the fair regularity of the doublet system; perhaps indeed it was only a hanger on, not a true member of the court; and in so far as the word 'satellite' conveys this impression it is misleading, for to-day the 'satellite' is recognised to be as much a component of the multiplet as the other two lines; it differs only in being fainter.

D. The FUNDAMENTAL series was formerly thought to consist of single lines, but the work which showed the presence of a satellite in the diffuse series, showed also that the fundamental series consists of doublets with a constant separation equal to that of the satellite and the chief line in the first diffuse pair.

From these facts certain consequences quite clearly follow. Thus if the sharp doublets have a constant interval, the ^{2}P term must be double, while the ^{2}S term varying as it does from doublet to doublet must be single. In the principal series on the other hand the intervals decrease, and accordingly the interval $m\,{}^2P_{1\frac{1}{2}}-m\,{}^2P_{\frac{1}{2}}$, often abbreviated to $\Delta^2P_{1\frac{1}{2},\frac{1}{2}}$ or even Δ^2P, must decrease as the serial number increases. Further the interval of the P terms is in general greater than that of the D terms, and the latter in turn greater than that of the F terms.

In symbols the doublets of the sharp series thus appear as

$$\left.\begin{matrix} m\,{}^2S_{\frac{1}{2}} \rightarrow 2\,{}^2P_{\frac{1}{2}} \\ m\,{}^2S_{\frac{1}{2}} \rightarrow 2\,{}^2P_{1\frac{1}{2}} \end{matrix}\right\}$$

with a constant separation of $2\,{}^2P_{1\frac{1}{2}} - 2\,{}^2P_{\frac{1}{2}}$; while the principal series doublets appear as

$$\left.\begin{matrix} m\,{}^2P_{\frac{1}{2}} \rightarrow 1\,{}^2S_{\frac{1}{2}} \\ m\,{}^2P_{1\frac{1}{2}} \rightarrow 1\,{}^2S_{\frac{1}{2}} \end{matrix}\right\}$$

the two lines approaching one another as m increases and having the same limit $1\,{}^2S_{\frac{1}{2}}$ (Fig. 3·7).

4. The spinning electron

The theory thus far outlined accounts both for a number of series, each having a characteristic value of l, and a number of terms in each series, the terms being determined by n. But it does not explain how these levels split into two or more components; to fill this gap Uhlenbeck and Goudsmit* suggested in 1925 that the electron itself may be supposed to spin with an angular momentum **s**, **s** being measured in the usual units $h/2\pi$ and being numerically equal to $\frac{1}{2}$. The angular momentum of the electron may then be represented in magnitude and direction by a vector **s**,

* Uhlenbeck and Goudsmit, *Nw*, 1925, **13** 953; *N*, 1926, **117** 264.

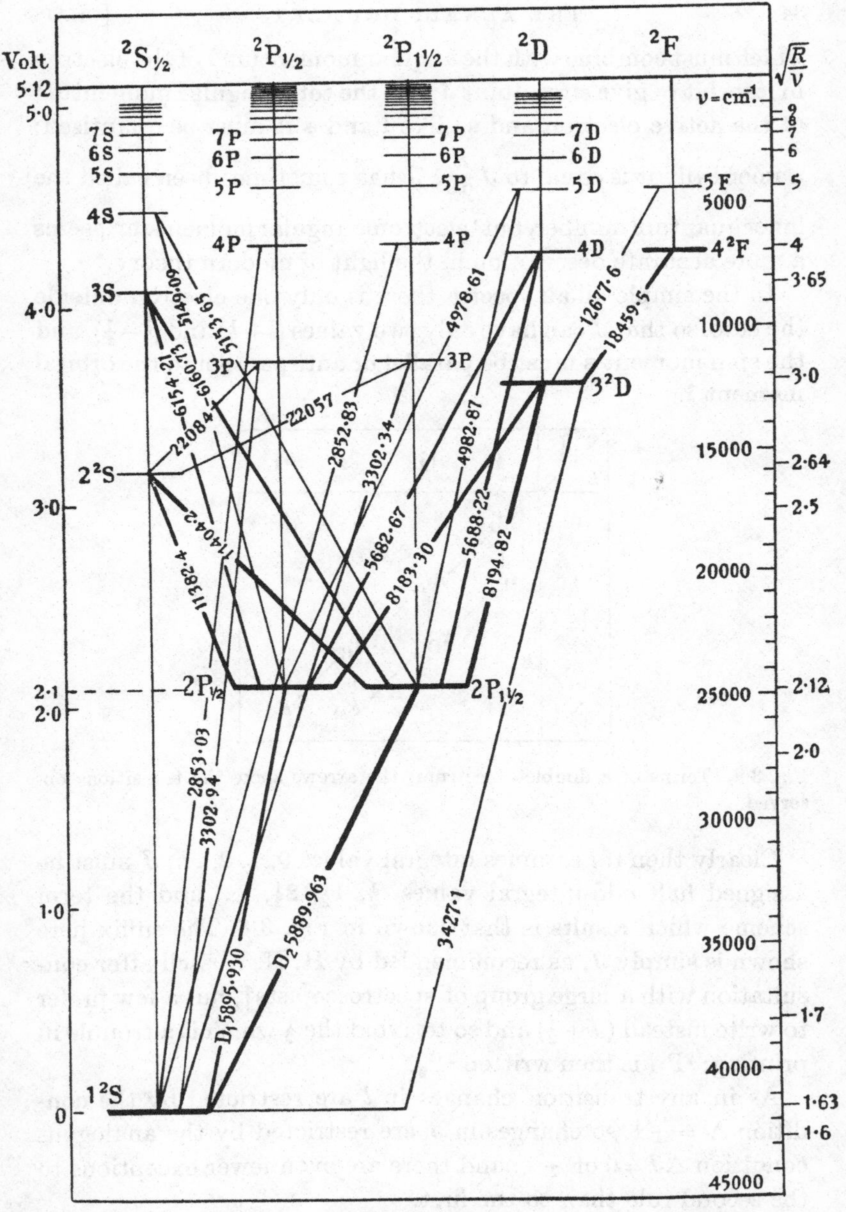

Fig. 3·7. Level diagram of sodium. Contrast this diagram with lithium (Fig. 3·3), in which not even the P levels are resolved, and with caesium (Fig. 3·15), in which the P, D and F levels all show doublet structure. After Grotrian, *Graphische Darstellung der Spektren.*

which must combine with the angular momentum **l** of the electron in its orbit to give a resultant **J**; **J** is the total angular momentum of the active electron and so like **l** and **s** it must be quantised; numerically it is equal to $J\frac{h}{2\pi}$. J has sometimes been called the inner quantum number, but 'electronic angular momentum' seems a more accurate description in the light of modern theory.*

In the simple alkali spectra there is only one electron outside the core, so that J can have only two values $(l+\frac{1}{2})$ and $(l-\frac{1}{2})$, and the spin moment **s** must be parallel or anti-parallel to the orbital moment **l**.

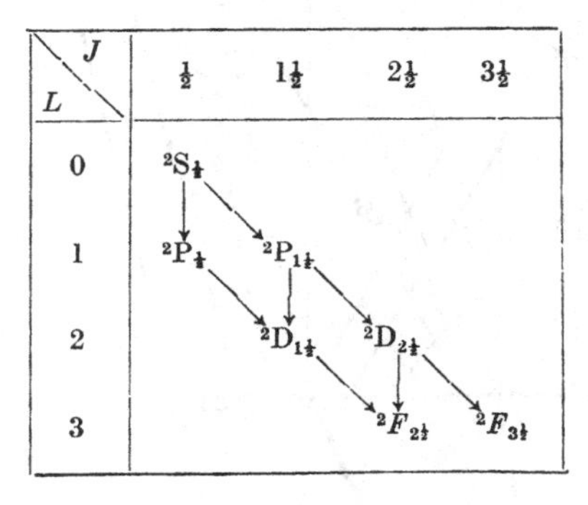

Fig. 3·8. Terms of a doublet spectrum; the arrows show the transitions observed.

Clearly then if l assumes integral values 0, 1, 2, ..., J must be assigned half odd-integral values, $\frac{1}{2}$, $1\frac{1}{2}$, $2\frac{1}{2}$, ...; and the term scheme which results is that shown in Fig. 3·8. The suffix here shown is simply J, as recommended by H. N. Russell after consultation with a large group of spectroscopists†, but a few prefer to write instead $(J+\frac{1}{2})$ and so to avoid the $\frac{1}{2}$ which is a trouble in printing; $^2P_{1\frac{1}{2}}$ is then written $^2P_2$.

As in any transition changes in l are restricted by the condition $\Delta l = \pm 1$, so changes in J are restricted by the analogous condition $\Delta J = 0$ or ± 1; and there are even fewer exceptions to the second rule than to the first.

* Goudsmit, *PR*, 1931, **37** 1501.

† Russell, Shenstone and Turner, *PR*, 1929, **33** 900.

Thus the three lines which should appear in the diffuse series of the alkalis are

$$\left.\begin{array}{l} m\,{}^2\mathrm{D}_{1\frac{1}{2}} \rightarrow 2\,{}^2\mathrm{P}_{\frac{1}{2}} \\ m\,{}^2\mathrm{D}_{1\frac{1}{2}} \rightarrow 2\,{}^2\mathrm{P}_{1\frac{1}{2}} \\ m\,{}^2\mathrm{D}_{2\frac{1}{2}} \rightarrow 2\,{}^2\mathrm{P}_{1\frac{1}{2}} \end{array}\right\} \begin{array}{l}\text{constant separation} \\ \\ \text{common limit}\end{array} \text{ identified as } \left\{\begin{array}{l}\alpha \\ \gamma \\ \beta\end{array}\right.$$

These formulae show that the lines of the first two series will have a constant separation equal to $2\,{}^2\mathrm{P}_{1\frac{1}{2}}-2\,{}^2\mathrm{P}_{\frac{1}{2}}$, while the second and third lines will approach one another as the serial number increases and will have a common limit $2\,{}^2\mathrm{P}_{1\frac{1}{2}}$.

These two facts suffice to match the lines of theory with the empirical lines, called α, β, γ in Fig. 3·6. The line which shares in the constant separation and the common limit is $m\,{}^2\mathrm{D}_{1\frac{1}{2}} \rightarrow {}^2\mathrm{P}_{1\frac{1}{2}}$, which may therefore be identified as γ; and the other two lines follow. Moreover, this identification is easily confirmed by comparing the intervals of the first line of the diffuse series with the constant intervals of the sharp and fundamental series.

Now in caesium α is of higher frequency than γ, so that ${}^2\mathrm{P}_{\frac{1}{2}}$ represents a state of less energy than ${}^2\mathrm{P}_{1\frac{1}{2}}$; and as the symbols are taken to denote atomic energies, ${}^2\mathrm{P}_{\frac{1}{2}} < {}^2\mathrm{P}_{1\frac{1}{2}}$. Similarly since β is of higher frequency than γ, ${}^2\mathrm{D}_{1\frac{1}{2}} < {}^2\mathrm{D}_{2\frac{1}{2}}$.

This arrangement with the component of greater J representing the greater energy is of very common occurrence in the alkalis, and is often referred to as 'regular'; but the description is not satisfactory, since in many elements the order is reversed. To avoid confusion, therefore, the doublet term of the alkalis will be described as 'erect', while when the component with the greater J has less energy, the term will be described as 'inverted'.

Applying this rule to the sharp series, the lines $m\,{}^2\mathrm{S}_{\frac{1}{2}} \rightarrow 2\,{}^2\mathrm{P}_{\frac{1}{2}}$ will be of greater frequency and shorter wave-length than the lines $m\,{}^2\mathrm{S}_{\frac{1}{2}} \rightarrow 2\,{}^2\mathrm{P}_{1\frac{1}{2}}$; while in the principal series the lines $m\,{}^2\mathrm{P}_{\frac{1}{2}} \rightarrow 1\,{}^2\mathrm{S}_{\frac{1}{2}}$ will have the longer wave-length. The truth of these statements can, however, be tested only when the intensity rules have been reviewed.

5. Sommerfeld's intensity rule

Not only the relative wave-length, but even the relative intensities of the lines of a multiplet may be deduced from the L

and J values of the levels concerned.* Sommerfeld has shown on the basis of the correspondence principle that the strongest lines occur when both L and J change in the same sense; these are sometimes referred to as the 'chief lines'. Weaker lines or 'satellites of the first order' occur when $\Delta J = 0$; while the weakest lines of all, 'satellites of the second order', result when L and J change

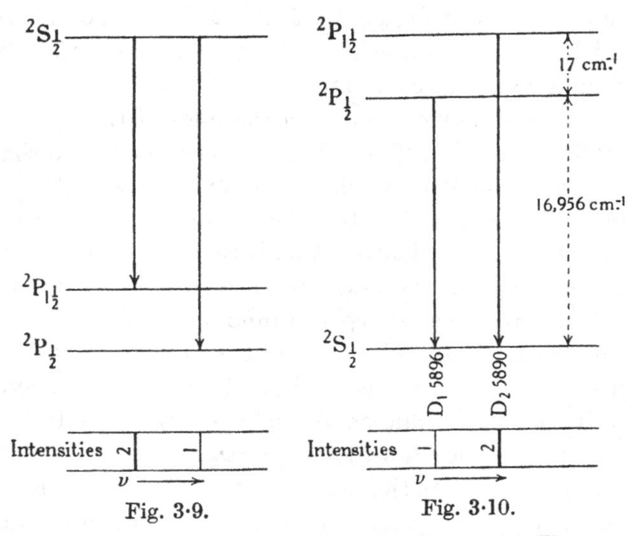

Fig. 3·9. Fig. 3·10.

Fig. 3·9. Transitions producing a doublet of the sharp series. The arrows are so spaced as to give the lines their proper positions in the frequency scale. The numbers beside the lines show the relative intensities.

Fig. 3·10. A doublet of the principal series. The intervals shown are those of the D_1D_2 doublet of sodium; they indicate the distortion introduced into this type of diagram.

in opposite senses. In a table of terms, such as that shown in Fig. 3·8, the three degrees of intensity correspond to the three directions in which the arrow may run; the sense of the arrow is of no account.

This work has been developed by other workers so that relative intensities of the components of a multiplet can be predicted, but the work involves heavy mathematics and Sommerfeld's rule is

* For the present L and S may be taken as identical with the quantities previously written l and s.

sufficient to justify certain predictions. In the sharp series, for instance, the line $m\,{}^2S_{\frac{1}{2}} \rightarrow 2\,{}^2P_{1\frac{1}{2}}$ with the lower frequency will be the stronger (Fig. 3·9); while in the principal series $m\,{}^2P_{1\frac{1}{2}} \rightarrow 1\,{}^2S_{\frac{1}{2}}$ will be the stronger, but this time the P term lies above the S term and accordingly it has a higher frequency than the other component $m\,{}^2P_{\frac{1}{2}} \rightarrow 1\,{}^2S_{\frac{1}{2}}$ (Fig. 3·10).

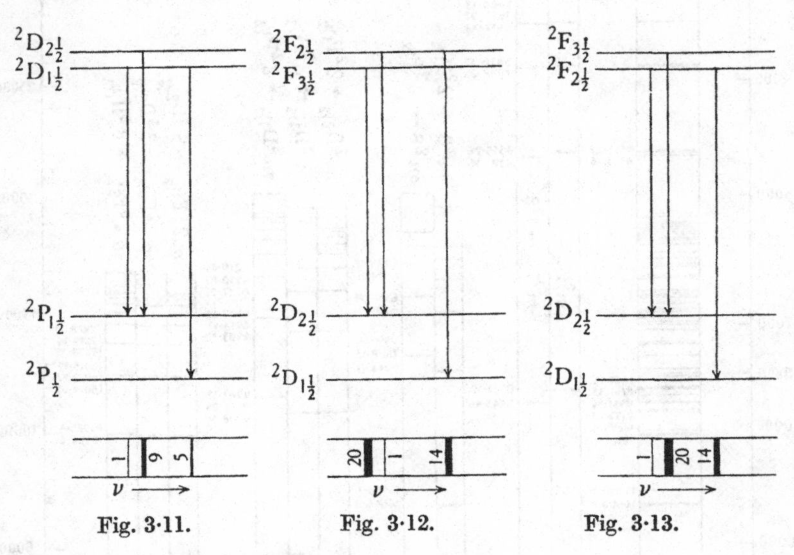

Fig. 3·11. **Fig. 3·12.** **Fig. 3·13.**

Fig. 3·11. Diffuse series doublet.

Fig. 3·12. Fundamental doublet; D term erect and F inverted.

Fig. 3·13. Fundamental doublet; both terms erect.

In the diffuse doublet the P levels have a greater interval than the D levels, so that where both terms are erect the relative position of the three lines will be that shown in Fig. 3·11; the intensity rule indicates the weakness of the satellite, but, unlike the more precise theory, it does not indicate what the relative intensity will be, nor even which of the chief lines is the stronger.

Theory suggests that the fundamental doublet like the diffuse doublet consists of three lines, but actually the two F levels are so close together that they have been resolved only in caesium (Figs. 3·14, 3·15); and in caesium the satellite lies not on the less refrangible side of the two chief lines but between them.* This may be taken to prove that the F levels of caesium are inverted, as

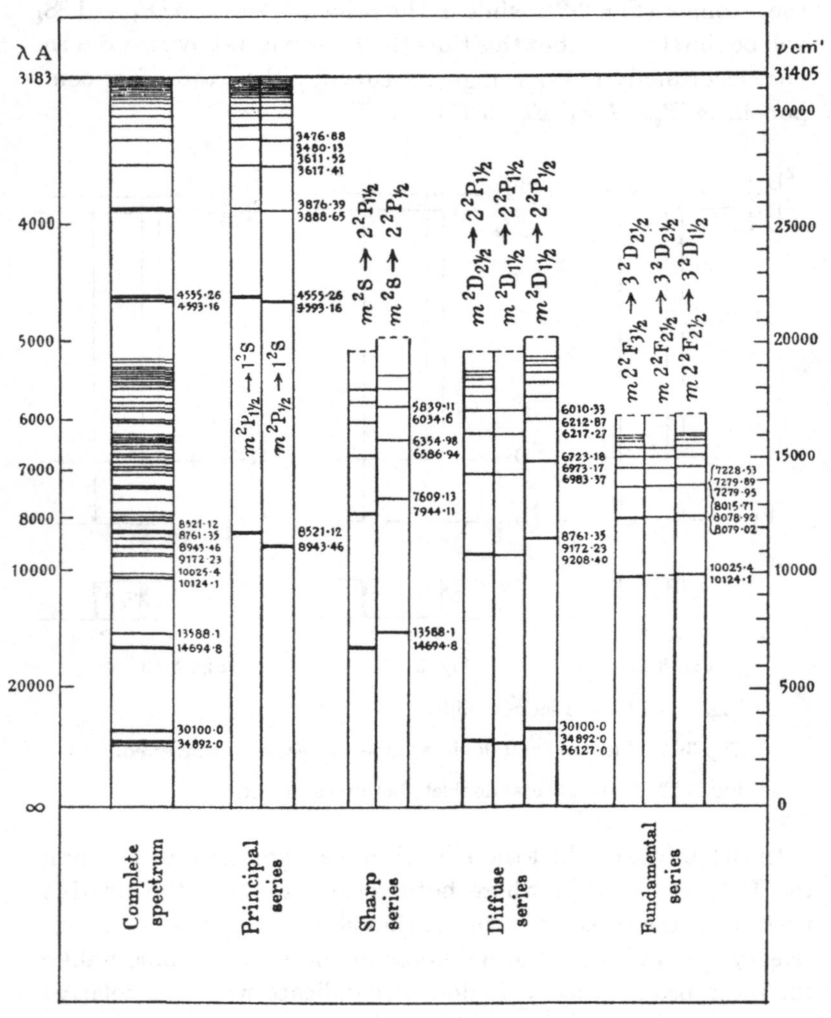

Fig. 3·14. Series in caesium.

shown in Fig. 3·12, for if both terms were erect the lines would be arranged like the diffuse doublet (Fig. 3·13).

* Meissner, *AP*, 1921, **65** 378.

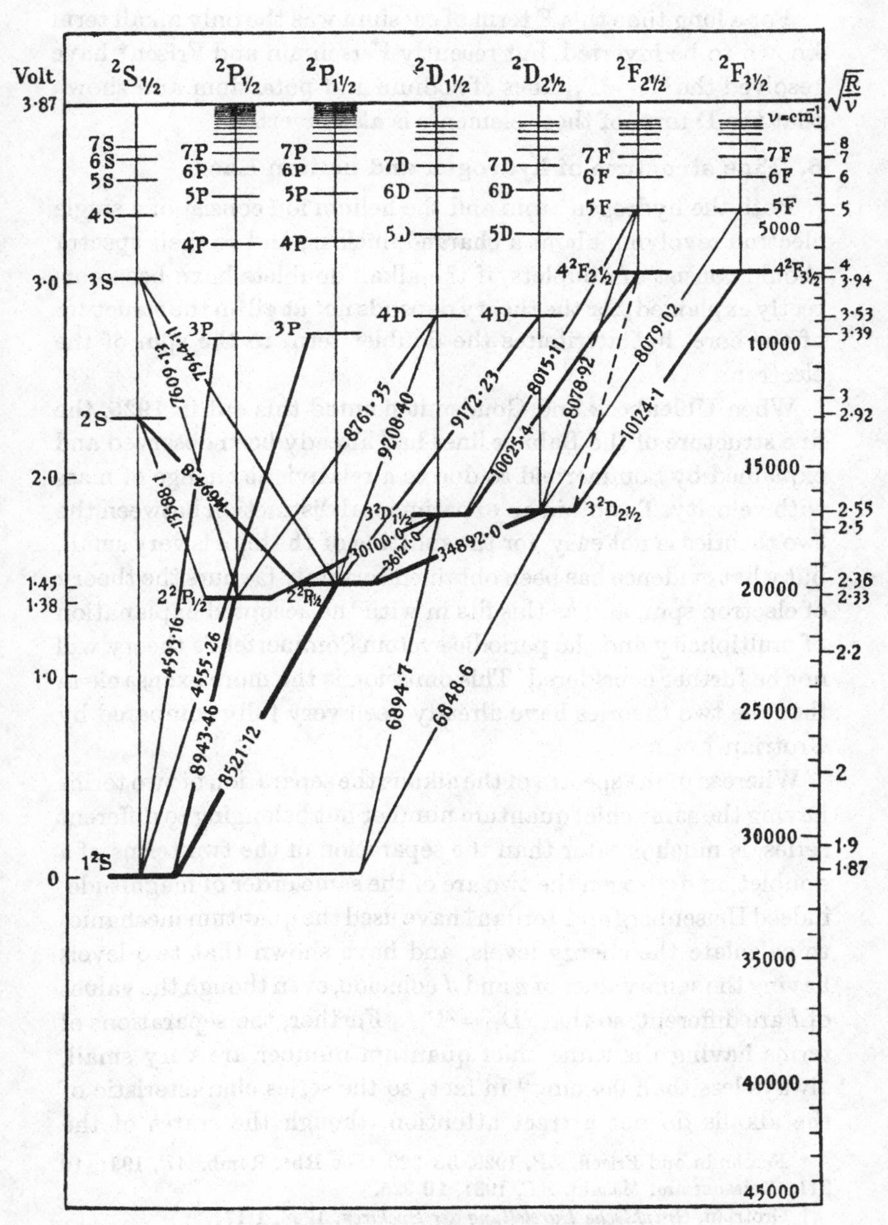

Fig. 3·15. Level diagram of caesium.

For a long time this F term of caesium was the only alkali term known to be inverted, but recently Ferschmin and Frisch* have resolved the $^2D \rightarrow {}^2P_{1\frac{1}{2}}$ lines of sodium and potassium and shown that the D term of these elements is also inverted.

6. Fine structure of hydrogen and helium lines

Both the hydrogen atom and the helium ion consist of a single electron revolving about a charged nucleus, and so their spectra should consist of doublets, if the alkali doublets have been correctly explained; for the theory depends not at all on the structure of the core, but attributes the doublet term to the spin of the electron.

When Uhlenbeck and Goudsmit pointed this out in 1925, the fine structure of the Balmer lines had already been observed and explained by Sommerfeld as due to a relativistic change of mass with velocity. To obtain an experimental distinction between the two theories is not easy, for the splitting of the lines is very small, but what evidence has been obtained certainly favours the theory of electron spin, and as this fits in with the accepted explanation of multiplicity and the periodic system Sommerfeld's theory will not be further considered. This omission is the more excusable in that the two theories have already been very fully compared by Grotrian.†

Whereas in the spectra of the alkalis the separation of two terms having the same chief quantum number but belonging to different series, is much greater than the separation of the two terms of a doublet, in hydrogen the two are of the same order of magnitude; indeed Heisenberg and Jordan‡ have used the quantum mechanics to calculate the energy levels, and have shown that two levels having the same values of n and J coincide, even though the values of l are different, so that $^2D_{1\frac{1}{2}} = {}^2P_{1\frac{1}{2}}$. Further, the separations of terms having the same chief quantum number are very small, always less than 0·4 cm.$^{-1}$ in fact; so the series characteristic of the alkalis do not attract attention, though the states of the

* Ferchmin and Frisch, *ZP*, 1929, **53** 326. For Rb I, Ramb, *AP*, 1931, **10** 311. Meissner and Masaki, *AP*, 1931, **10** 325.

† Grotrian, *Graphische Darstellung der Spektren*, 1928, **1** 17f.

‡ Heisenberg and Jordan, *ZP*, 1926, **37** 263.

electron are still determined by n, l and J, and the same term scheme is therefore valid.

These predictions have been experimentally tested only on three Balmer lines in the visible spectrum H_α, H_β and H_γ. Of these H_α, corresponding to the transition $n=3 \rightarrow n=2$, is shown diagrammatically in Fig. 3·17. On the left of the levels in this figure is set the modern doublet notation and on the right an earlier notation used by Sommerfeld. Five transitions are allowed, but of these two are so much stronger than the other three that H_α was long considered to be a doublet. The size of this doublet has been measured three times, and all workers agree on a value of about 0·318 cm.$^{-1}$ or 0·010 cm.$^{-1}$ less than the theoretical value of 0·328 cm.$^{-1}$ (Fig. 3·16).

Line	$\Delta\nu$ theory cm.$^{-1}$	$\Delta\nu$ observed. Hansen[a]	$\Delta\nu$ observed. Houston[b]	$\Delta\nu$ observed. Kent, Taylor and Pearson[c]
H_α	0·328	0·316	0·315	0·318
H_β	0·349	0·317	0·331	0·335
H_γ	0·357	0·328	0·353	0·354

Fig. 3·16. Doublet separations of three Balmer lines. The theoretical interval is the separation of the two strongest components; the observed interval is that of the empirical doublet.

[a] Hansen, *AP*, 1925, **78** 558.
[b] Houston, *AJ*, 1926, **64** 81.
[c] Kent, Taylor, L. B. and Pearson, *PR*, 1927, **30** 266.

Recognising that this discrepancy was greater than the probable error, Hansen examined the photometric curve reproduced in Fig. 3·18 and showed that the observed intensity distribution could be explained by the addition of three curves, each having the form of an error-curve e^{-x^2}. He chose the intensities, positions and half-value breadth of the components so as to obtain agreement with the empirical curve at the two maxima and at the position of a third component. A certain additional correction is really needed because the Lummer plate used to obtain the photographs always broadens the base of the curve, but even ignoring this, the agreement is as good as can be expected. Particularly satisfactory is Hansen's measurement of the separation of the two chief components as 0·326 cm.$^{-1}$, which is now in

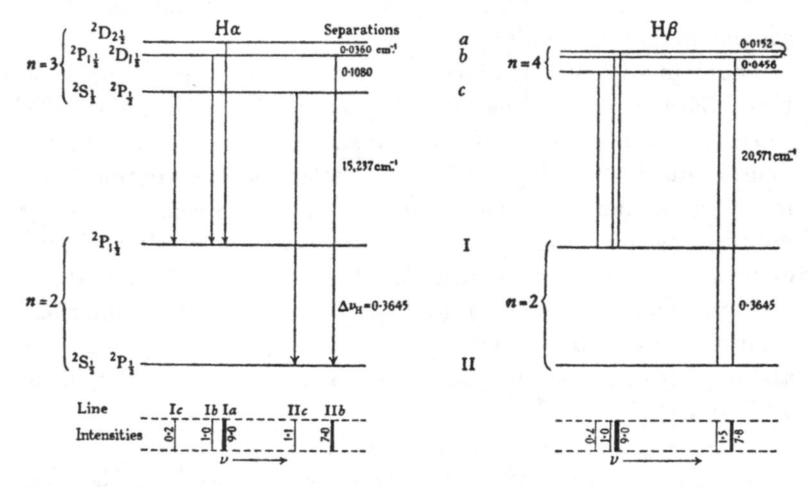

Fig. 3·17. Theoretical structure of the first two lines of the Balmer series, H_α and H_β.

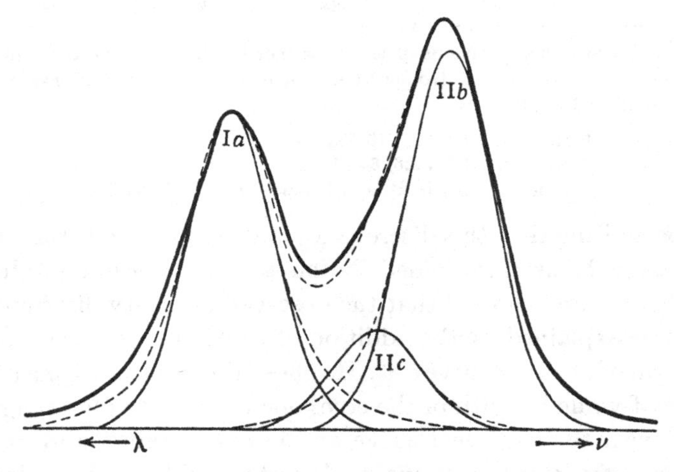

Fig. 3·18. Intensity distribution in H_α after Hansen, *AP*, 1925, **78** 558. The heavy line is a microphotometer trace. The light curves are the three error curves which Hansen chose so as to obtain agreement, at the two maxima and at the position of a third component. The lower dotted curves show the broadening of the theoretical curves, which the Lummer plate used would produce; the upper dotted curve is obtained by summing the lower curves.

satisfactory agreement with theory. The only outstanding difficulty is that theory would make the long wave-length component I*a* the stronger of the two strong lines, while Hansen's measurements show it as the weaker. Sommerfeld's theory, though agreeing with experiment in every other point, cannot account for the II*c* component or for any component lying between the bright pair II*b* and I*a*.

In passing from H_α to H_β the separation of levels I and II remains unchanged, but *a*, *b* and *c* approach one another so that the observed separation of the apparent doublet II*b* and I*a* will increase; this is shown quite clearly in Fig. 3·17, and is confirmed by the measurements recorded in Fig. 3·16. This change is carried a step further in H_γ, and again experimental confirmation is satisfactory.

Probably this agreement would have satisfied physicists for many years to come, had not heavy hydrogen been discovered. This led, however, to renewed investigation for two reasons; first the separation of the H_α line of H^1 and H^2 offers an accurate method of comparing the masses of the electron and atom*; and second one naturally asks whether the structure of the two lines is identical. It seems they are; but while confirming this point, Spedding, Shane and Grace† confirmed Hansen's statement that the intensities do not agree with theory; whereas the intensities of the three strongest components should be in the ratio of 9·00 : 7·08 : 1·13, they are in the ratio of 6·86 : 7·08 : 1·78, the second line being taken as standard. The deviations can be explained qualitatively if the $3\,^2D$ level carries less atoms than it would do in the thermodynamic equilibrium postulated by theory.

As the theoretical predictions of Uhlenbeck and Goudsmit have been confirmed by measurements on the hydrogen lines, so also they explain the structure of the 4686 A. line of ionised helium. It is the first line of the Fowler series and corresponds to the transition $n=4 \rightarrow n=3$, consequently the fine structure should be

* Shane and Spedding, *PR*, 1935, **47** 33.

† Spedding, Shane and Grace, *PR*, 1935, **47** 38. Cf. Houston and Hsieh, *PR*, 1934, **45** 263. Williams and Gibbs, *PR*, 1934, **45** 475. These papers are not in entire agreement.

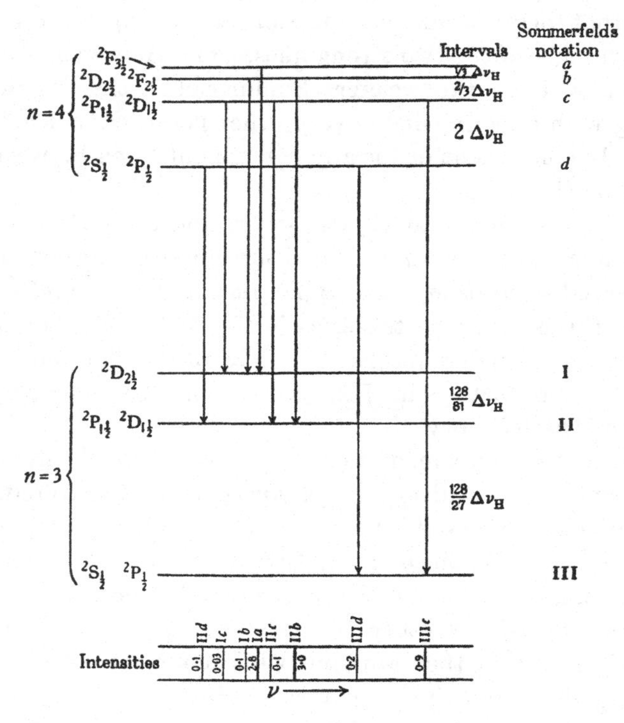

Fig. 3·19. Transitions producing the 4686 A. line of He II.

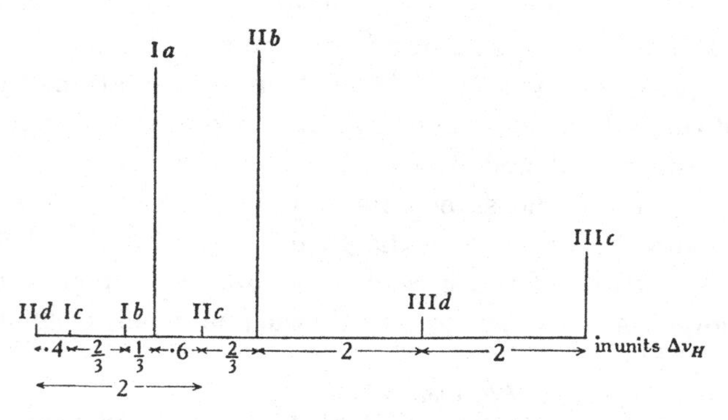

Fig. 3·20. Theoretical structure of the 4686 A. line of He II.

that shown in the Grotrian diagram of Fig. 3·19. The separations of this figure and the intensities calculated by Sommerfeld and Unsöld,* agree with Paschen's† photometric curve (Fig. 3·21).

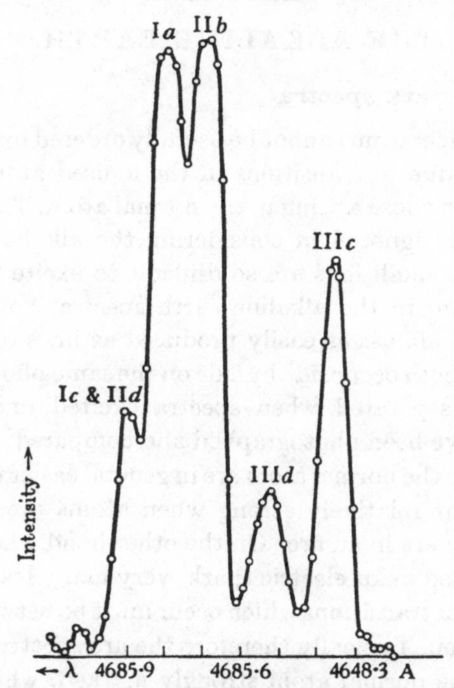

Fig. 3·21. Microphotometer curve of the 4686 A. line of He II after Paschen, *AP*, 1927, **82** 689.

As in hydrogen so also in ionised helium the older theory of Sommerfeld succeeds marvellously well almost to the end, but it cannot account for the III*d* component and gives the relative intensities of the two strong components wrong.

BIBLIOGRAPHY

The most thorough account of the alkali doublets is Grotrian, *Graphische Darstellung der Spektren von Atomen mit ein, zwei und drei Valenzelectronen*, 1928, **1**. The second volume, containing level diagrams and the wave-length of the line arising in each transition, is invaluable.

* Sommerfeld and Unsöld, *ZP*, 1926, **36** 259; **38** 237. The separations can be calculated from equation (11·6) of this book.

† Paschen, *AP*, 1927, **82** 689.

CHAPTER IV

THE ALKALINE EARTHS

1. Arc and spark spectra

The lines of a spectrum cannot be usefully ordered into series until any that are due to transitions in the ionised atom have been separated from those arising in the normal atom. This point may be legitimately ignored in considering the alkalis because the spectra of the alkali ions are so difficult to excite that there is little confusion; in the alkaline earth spectra, however, many ionic lines are almost as easily produced as lines of the normal atom, so that both occur side by side on the same photograph, and can only be separated when spectra excited under different conditions have been photographed and compared.

Lines due to the normal atom are in general easier to excite, and so they appear relatively strong when atoms are slightly disturbed as they are in an arc. On the other hand when atoms are roughly handled in an electric spark, very many lose an electron and any further transitions which occur must be between different states of the ion. Typically therefore the arc spectrum will show the lines of the normal atom strongly marked, while the spark will bring out or 'enhance' lines due to the ion; and so the former are commonly referred to as 'arc lines' and the latter as 'spark lines', while lines due to an atom which has lost two electrons are sometimes said to belong to the 'second spark spectrum'.

There are other ways in which the two spectra may be separated. Thus King has measured the intensities of lines in an oven maintained at a steady temperature, and repeating the work for other temperatures has found how the intensities vary; typically spark lines develop only at higher temperatures. Or again potassium may be added to the furnace, for being easily ionised it depresses the ionisation of the substance under examination, according to the chemical law of mass action, and so diminishes the intensity of any lines due to the ion. In an arc too lines due to

PLATE II

1. Arc and spark spectra of calcium compared. The term written above each line is the first term of the transition. The doublet lines which arise in the calcium ion are brighter in the spark spectrum, but the singlet and triplet lines which arise in the normal atom are brighter in the arc. The shorter lines in the spark spectrum are due to nitrogen and oxygen.

2. A fundamental triplet of barium, showing how diffuse are the lines of the arc in air, and the fine structure of these lines when seen in the electric furnace at low pressure. (A. S. King, Mount Wilson Observatory.)

3. Multiplets in scandium. Whereas the doublet lines which arise in the normal atom stretch right across the spectrum, the triplet lines which arise in the ion are brighter near the positive pole of the arc. The band down the centre is an iron arc used for calibration. The lines shown stretch from 3075 A. on the left to 2981 A. on the right. A cross denotes a singlet line.

4. Multiplets in scandium. The $b\,{}^4F^\circ \rightarrow a\,{}^4F$ multiplet stretches from 5101 A. to 5064 A., while the singlet line is 5031 A. The grouping of the strong lines arising in $J \rightarrow J$ transitions is characteristic of an $L \rightarrow L$ multiplet.

(Figs. 1 and 2 were lent by Prof. A. Fowler. Figs. 3 and 4 by Dr W. F. Meggers.)

Plate II

1. Calcium

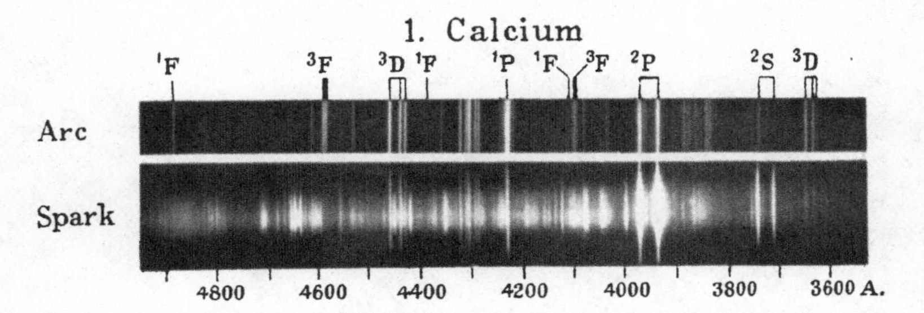

2. Fundamental Triplet of Barium

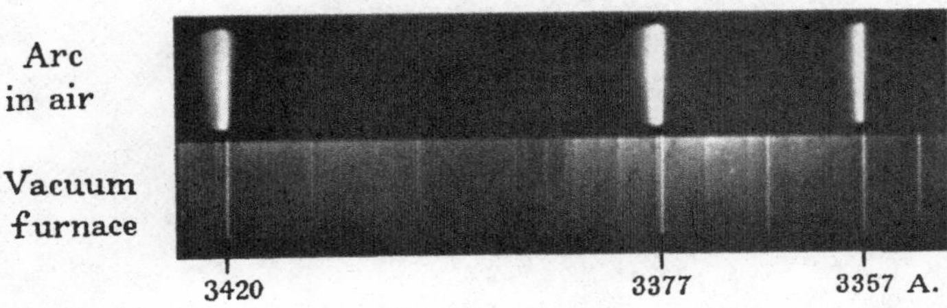

3. Multiplets in Scandium

$a\,^3G_3$

$a\,^3G_4$

$a\,^3G_5$

$a\,^3F^\circ_4$

$a\,^3F^\circ_3$

$a\,^3F^\circ_2$

$a\,^2D_{2\frac{1}{2}}$

$a\,^2D_{1\frac{1}{2}}$

$c\,^2F^\circ_{2\frac{1}{2}}$

$c\,^2F^\circ_{3\frac{1}{2}}$

$c\,^2D^\circ_{2\frac{1}{2}}$

$c\,^2D^\circ_{1\frac{1}{2}}$

4. Multiplets in Scandium

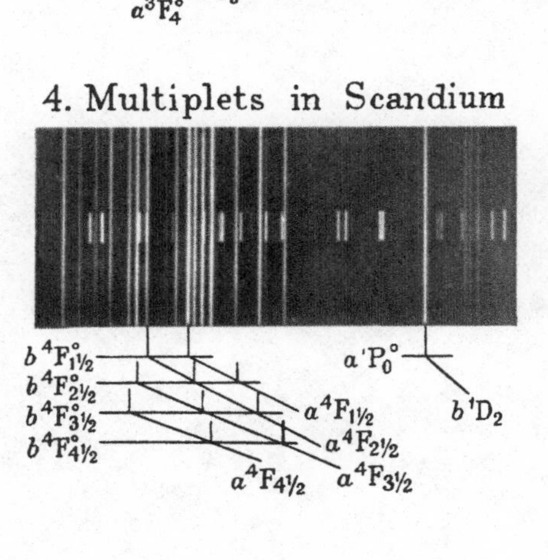

an ion will appear particularly strong near the negative electrode, whither the positive ions are drawn.

The result is a large number of lines all arising from a definite atom in a definite state of ionisation. This is the rock on which any theory is founded.

2. Two systems. The singlets

The spectra of the different elements of the second column of the periodic table resemble one another as did the spectra of the alkali metals. Each spectrum consists of a system of singlets and a system of triplets, and the separations of the latter increase roughly as the square of the atomic number (Fig. 4·1).

In each system there are found four series which are again known as principal, sharp, diffuse and fundamental; the sharp and diffuse series still have the same limit; and there can be no doubt that in the four term sequences, the orbital quantum number L has the values 0, 1, 2 and 3, for the S terms combine only with the P terms as before, while P, D and F terms combine with the two sequences allowed by the selection rule $\Delta L = \pm 1$. Thus the old series formulae are still valid, and there is only one important difference, the triplet lines most easily obtained in absorption belong not to the principal but to the sharp and diffuse series.

The doublets of the alkalis were explained by endowing the electron with an angular momentum of $\frac{1}{2}$; so to account for the two systems of the alkaline earths with as few new assumptions as possible, assume that each electron still has a spin moment of $\frac{1}{2}$; then if these angular momenta $\mathbf{s}_1$, $\mathbf{s}_2$ combine to a resultant $\mathbf{S}$, and this resultant is itself subject to quantum conditions, the only possible values for S are 0 or 1 in units $h/2\pi$. Each electron will also have an angular momentum $\mathbf{l}$ due to its motion in its orbit, but in the simple spectra here considered one of the two electrons has $l=0$, so that the resultant orbital vector $\mathbf{L}$ is identical with the electronic orbital vector $\mathbf{l}$ of the one active electron.

In the notation here adopted electronic quantities are written in small letters, and atomic quantities in capitals. In the discussion of the alkalis it is true that $\mathbf{l}$ and $\mathbf{s}$ were used where $\mathbf{L}$ and $\mathbf{S}$

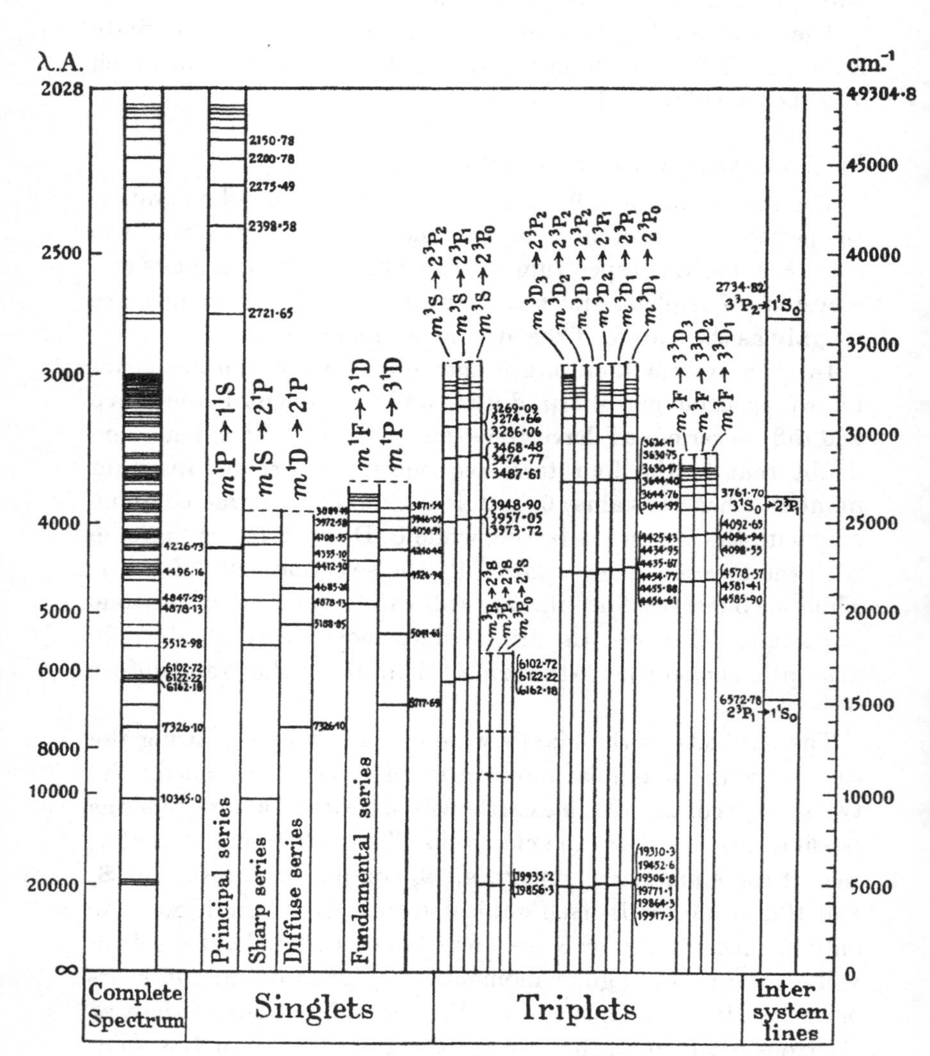

Fig. 4·1. Series in the arc spectrum of calcium. Note how much more complex this spectrum is than lithium (Fig. 3·1).

might have been written, but where only one electron is active small letters and capitals are synonymous.

The atomic spin vector **S** combines with the orbital vector **L** of the single excited electron, so that $\mathbf{L}+\mathbf{S}=\mathbf{J}$ and if the arithmetical value of **J** is $J\frac{h}{2\pi}$, then J will be determined by the condition

$$|L+S| \geqslant J \geqslant |L-S|,$$

where this is understood to mean that J assumes the two terminal values and a series of intermediate values differing by unity.

L \ J	0	1	2	3
0	^{1}S			
1		^{1}P		
2			^{1}D	
3				^{1}F

Fig. 4·2. Terms of a singlet spectrum.

L \ J	0	1	2	3	4
0		$^3S_1$			
1	$^3P_0$	$^3P_1$	$^3P_2$		
2		$^3D_1$	$^3D_2$	$^3D_3$	
3			$^3F_2$	$^3F_3$	$^3F_4$

Fig. 4·3. Terms of a triplet spectrum.

Thus when $S=0$, $J=L$, and the possible terms are shown in Fig. 4·2; this is the singlet system.

When $S=1$, however, $J=(L-1)$, L, or $(L+1)$ provided $L \geqslant 1$; but when $L=0$, since J must be positive, it can only have the value 1. The resulting scheme of terms is shown in Fig. 4·3, and represents the triplet system.

The singlet system is so simple that the Grotrian diagram of Fig. 4·4 is really sufficient description, but the allotment of the serial numbers deserves a word of explanation. In lithium the lowest terms of the four sequences were written 1S, 2P, 3D and 4F, and this convention will be retained, even though the deepest D terms of the singlet system of calcium lies below the deepest P terms and has an effective quantum number of 2; this term is therefore called 3 ^{1}D.

3. The triplet system

In discussing the doublets of the alkalis their appearance was first described, and certain rules were then developed to account

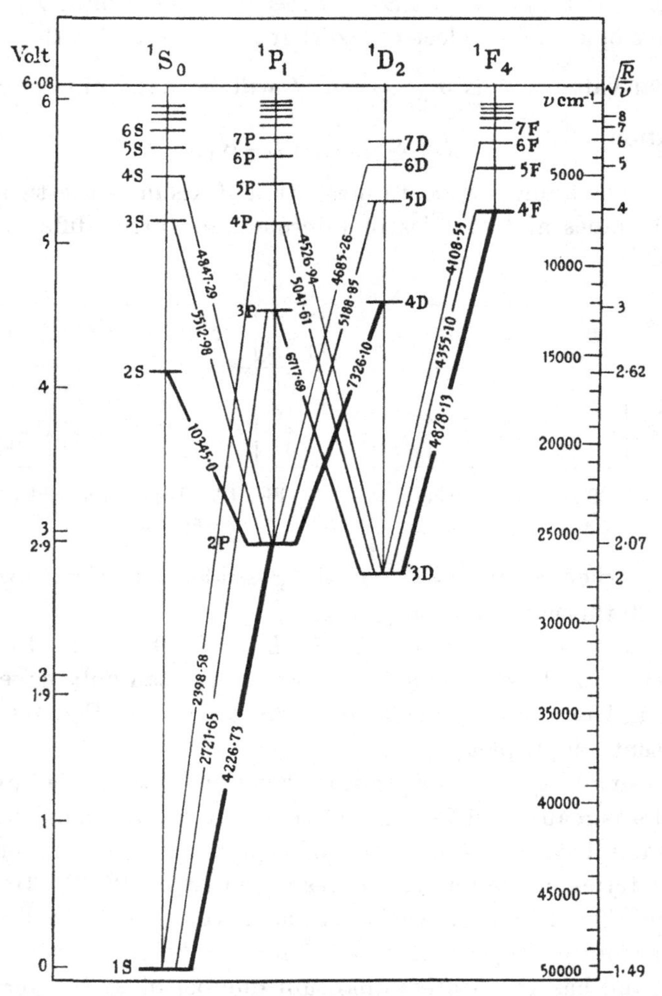

Fig. 4·4. Level diagram showing the singlet system of calcium. After Grotrian, *Graphische Darstellung der Spektren.*

for their structure. In the alkaline earths the procedure will be reversed; the rules developed will be applied to predict the structure and this will then be compared with experiment.

The rules do not serve to determine the relative term values, so these must first be given. The effective quantum numbers of the triplets of calcium are shown in Fig. 4·5 together with the corresponding serial numbers.

Serial number	Term series							
	^{1}S	^{1}P	^{1}D	^{1}F	^{3}S	^{3}P	^{3}D	^{3}F
1	1·49							
2	2·62	2·07			2·49	1·80		
3	3·82	2·95	2·00		3·52	2·93	1·95	
4	4·67	3·78	3·02	3·97	4·54	4·02	3·08	3·92
5	5·67	4·52	4·14	4·94	5·55	5·03	4·09	4·91

Fig. 4·5. Effective quantum numbers of calcium.

The lowest ^{3}D term has $n^\star = 1{\cdot}95$ and yet we write consistently with the above plan $m = 3$. On the other hand, we are apparently inconsistent in writing $m = 2$ in the deepest ^{3}S term; yet this notation is to be desired because it avoids an apparent exception to the rule, which states that the triplet term always lies slightly deeper than the corresponding singlet term; later, too, Pauli's exclusion principle will be seen to give theoretical ground for the hypothesis that the $1\,^3$S term is missing in alkaline earth spectra.

The Grotrian diagrams of calcium and mercury are shown in Figs. 4·4, 6 and 4·7. The ground state of the atom is $1\,^1$S, but the deepest of the triplet states is $2\,^3$P, which accounts for the appearance of the sharp and diffuse series in absorption.

The terms are erect, so that a typical triplet of the principal series will consist of three lines:

$m\,^3P_2 \rightarrow 2\,^3S_1$	Brightest	Highest ν	Shortest λ
$m\,^3P_1 \rightarrow 2\,^3S_1$			
$m\,^3P_0 \rightarrow 2\,^3S_1$	Faintest	Lowest ν	Longest λ

Thus intensity decreases with increasing wave-length. This is shown diagrammatically in Fig. 4·8, where the relative intensities are written alongside the lines. The separation decreases as the serial number increases and all three series have the same limit. In calcium experiment has revealed only the first term of this

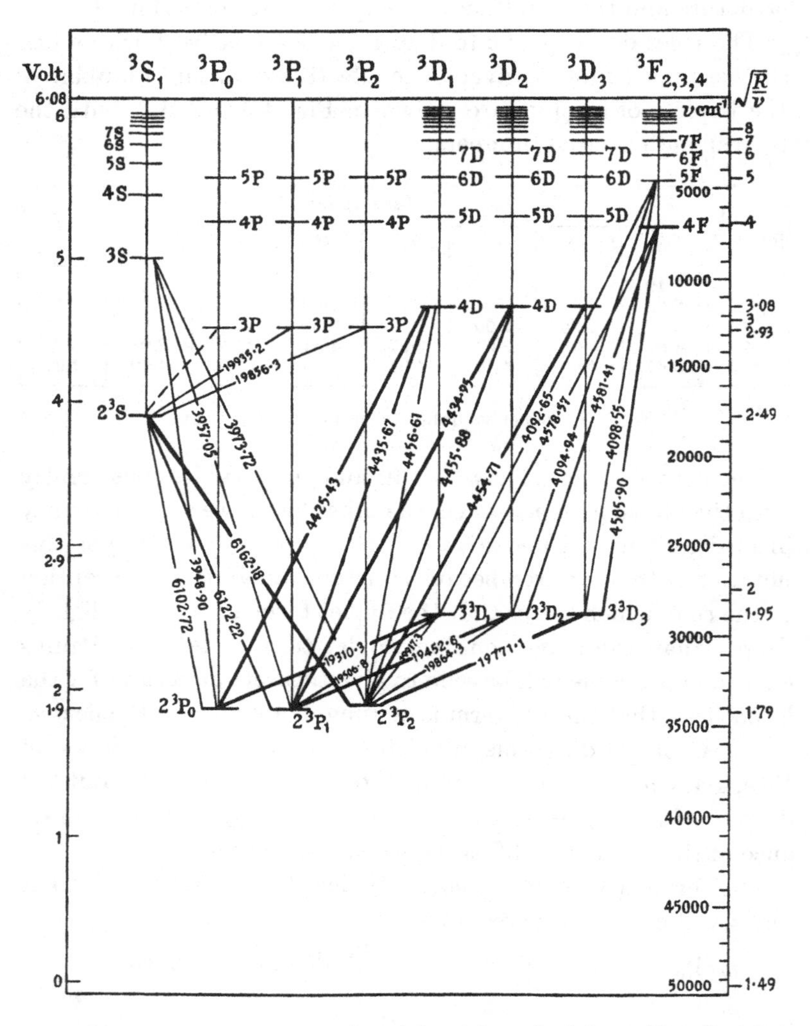

Fig. 4·6. Level diagram showing the triplet system of calcium. After Grotrian, *Graphische Darstellung der Spektren.*

series, because the $2\,^3S_1$ term lies so high, but in zinc several triplets are known.

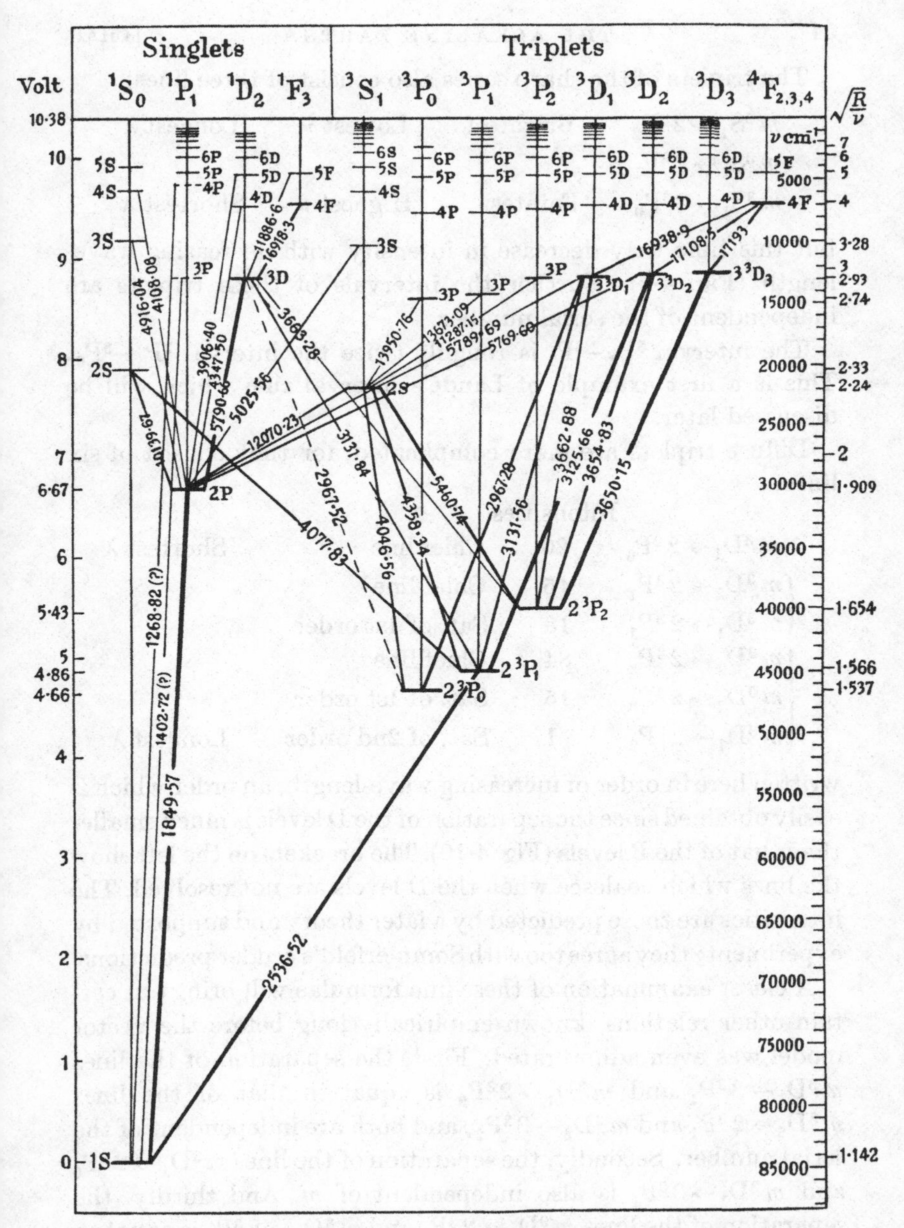

Fig. 4·7. Level diagram of mercury. After Grotrian, *Graphische Darstellung der Spektren*.

The triplets of the sharp series also consist of three lines:

$m\,^3S_1 \rightarrow 2\,^3P_2$	Brightest	Lowest ν	Longest λ
$m\,^3S_1 \rightarrow 2\,^3P_1$			
$m\,^3S_1 \rightarrow 2\,^3P_0$	Faintest	Highest ν	Shortest λ

but this time they decrease in intensity with decreasing wave-length (Fig. 4·9). Further the intervals of these triplets are independent of the serial number.

The interval $^3P_2 - {}^3P_1$ is roughly twice the interval $^3P_1 - {}^3P_0$. This is a first example of Landé's interval rule which will be discussed later.

Diffuse triplets are more complicated, for they consist of six lines:

	Intensities		
$m\,^3D_1 \rightarrow 2\,^3P_0$	20	Chief line	Shortest λ
{ $m\,^3D_2 \rightarrow 2\,^3P_1$	45	Chief line	
{ $m\,^3D_1 \rightarrow 2\,^3P_1$	15	Sat. of 1st order	
{ $m\,^3D_3 \rightarrow 2\,^3P_2$	84	Chief line	
{ $m\,^3D_2 \rightarrow 2\,^3P_2$	15	Sat. of 1st order	
{ $m\,^3D_1 \rightarrow 2\,^3P_2$	1	Sat. of 2nd order	Longest λ

written here in order of increasing wave-length, an order which is easily obtained since the separation of the D levels is much smaller than that of the P levels (Fig. 4·10). The brackets on the left show the lines which coalesce when the D levels are not resolved. The intensities are those predicted by a later theory and supported by experiment; they agree too with Sommerfeld's cruder predictions.

A closer examination of these line formulae will bring out certain other relations, known empirically long before the vector model was even adumbrated. First, the separation of the lines $m\,^3D_1 \rightarrow 2\,^3P_1$ and $m\,^3D_1 \rightarrow 2\,^3P_2$ is equal to that of the lines $m\,^3D_2 \rightarrow 2\,^3P_1$ and $m\,^3D_2 \rightarrow 2\,^3P_2$, and both are independent of the serial number. Secondly, the separation of the lines $m\,^3D_1 \rightarrow 2\,^3P_0$ and $m\,^3D_1 \rightarrow 2\,^3P_1$ is also independent of m. And thirdly, the separation of the lines $m\,^3D_1 \rightarrow 2\,^3P_1$ and $m\,^3D_2 \rightarrow 2\,^3P_1$ is equal to that of the $m\,^3D_1 \rightarrow 2\,^3P_2$ and $m\,^3D_2 \rightarrow 2\,^3P_2$, but the separation decreases as the serial number increases.

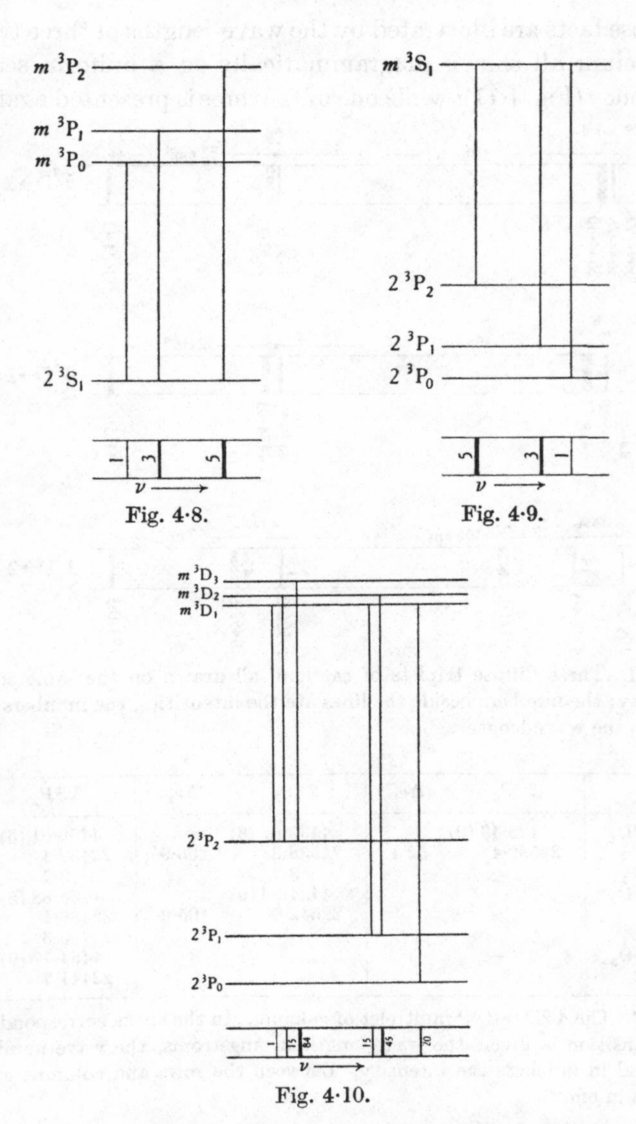

Fig. 4·8.

Fig. 4·9.

Fig. 4·10.

Fig. 4·8. Transitions producing a triplet of the principal series. The verticals are spaced so that the lines occupy their proper positions in the frequency scale; the numbers beside the lines show the relative intensities.

Fig. 4·9. A triplet of the sharp series.

Fig. 4·10. A diffuse triplet.

These facts are illustrated by the wave-lengths of three triplets of calcium all shown diagrammatically on a uniform scale of frequency (Fig. 4·11); while one of the three is presented again in a

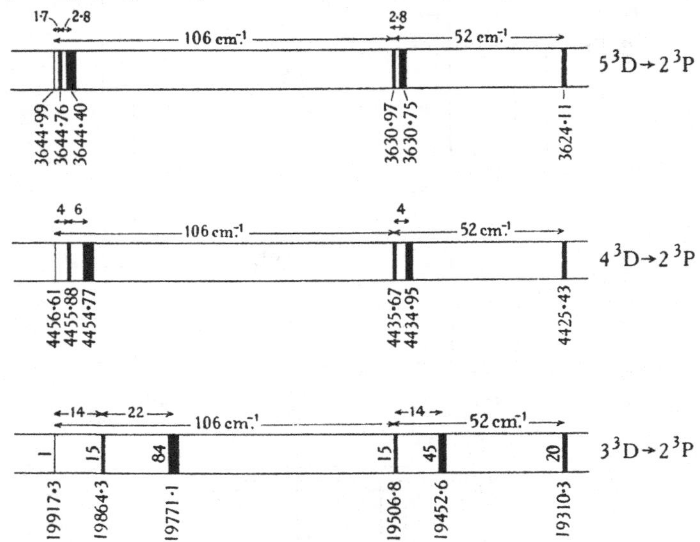

Fig. 4·11. Three diffuse triplets of calcium, all drawn on the same scale of frequency; the numbers beside the lines are the intensities, the numbers below the lines the wave-lengths.

	$2\,^3P_0$	$\Delta\nu_1$	$2\,^3P_1$	$\Delta\nu_2$	$2\,^3P_2$
$4\,^3D_1$	4425·43 (9) 22590·4	52·1	4435·67 (8) 22538·3	105·9	4456·61 (3) 22432·4
$\delta\nu_1$			3·7		3·7
$4\,^3D_2$	—		4434·95 (9) 22542·0	105·9	4455·88 (5) 22436·1
$\delta\nu_2$					5·6
$4\,^3D_3$	—		—		4454·77 (9) 22441·7

Fig. 4·12. The $4\,^3D \rightarrow 2\,^3P$ multiplet of calcium. In the space corresponding to each transition is given the wave-length in angstroms, the wave-number in cm.$^{-1}$ and in brackets the intensity. Between the rows and columns are the intervals in cm.$^{-1}$

form which is important in Fig. 4·12. The combining terms appear above and to the left, while, in the space corresponding to two terms, are written the wave-length and wave-number of the line resulting from the combination. Thus Fig. 4·12 shows that the

$4\,^3D_1 \rightarrow 2\,^3P_0$ line has a wave-length of 4425·43 A. and a wave-number of 22590·4 cm.$^{-1}$ Between the rows and columns are written the difference of the two wave-numbers, and these should of course be constant in any row or column, for they are the separations of the terms. To the right of the wave-length is sometimes written in a bracket the intensity of the line.

The complete fundamental triplet is not found in calcium, where the ^3F level has not been resolved, but appears in strontium and barium. The structure so closely resembles that of the diffuse triplet that a verbal description is hardly warranted, but the origin and intensity of the six lines is shown in Fig. 4·13.

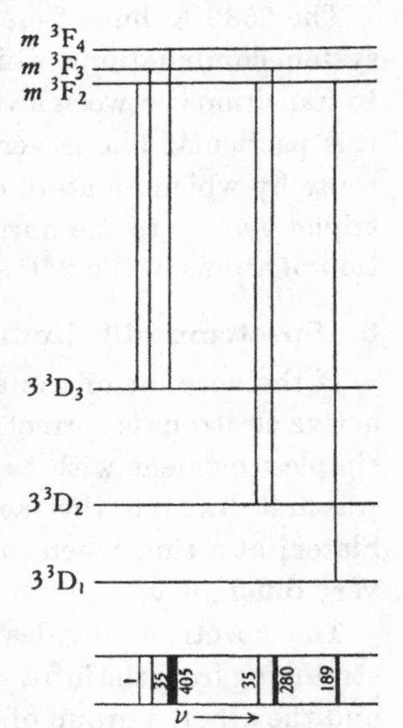

Fig. 4·13. Transitions producing a fundamental triplet.

The separation and intensities of all the lines of the triplet spectrum of calcium are thus easily derived from the term scheme by these simple rules. But the singlet and triplet system account for only half the bright lines of calcium; the other half arise from the simultaneous jumps of two electrons, as will appear when the displaced terms are discussed.

4. Inter-system lines

Besides the lines of the singlet and triplet systems, there are others which arise from the combination of singlet terms with triplet terms. These are governed by the selection rules already given; that is, an electron can jump from one state to another only when $\Delta L = \pm 1$ and $\Delta J = 0$ or ± 1. But to account for the absence of certain lines Landé* pointed out that an additional prohibition must be introduced, namely that an electron cannot jump from one state having $J=0$ to another also having $J=0$.

* Landé, *PZ*, 1921, **22** 417.

In the mercury spectrum, for instance, the line $2\,^3P_1 \rightarrow 1\,^1S_0$ at 2536 A. is very bright, but the line $2\,^3P_0 \rightarrow 1\,^1S_0$ is not ordinarily observed at all.

The 2536 A. line of mercury is not, however, typical of inter-system combinations, which are in general weaker than lines due to transitions between levels of the same system. On the contrary this particular line is very bright because it indicates the only route by which an atom can return from the lowest state of the triplet system to the normal state; consequently the concentration of atoms in the $2\,^3P$ state is often very high.

5. Spectrum of helium*

If the account of the alkaline earth spectra in terms of two active electrons is correct, the theory should be applicable to the simplest element with two electrons, namely helium. Attention was first drawn to this fact by Goudsmit and Uhlenbeck† and by Slater‡ at a time when the spectrum was commonly interpreted very differently.

The spectrum divides itself naturally into two parts, one stretching from the infra-red through the visible to about 2600 A., and the other, a group of half a dozen lines in the extreme ultra-violet, all with wave-lengths of less than 600 A. (Fig. 4·14). The visible and near visible regions were analysed by Runge and Paschen as long ago as 1896, but not until 1922 did Lyman§ discover the lines in the extreme ultra-violet. The term scheme, worked out by Runge and Paschen, consisted of two systems, one of singlets and the other of doublets; each system consisted quite normally of a principal, sharp, diffuse and fundamental series, but no connection between the two could be found; and as two unrelated systems were not known in any other element, it was suggested that they arose from two different atoms, called orthohelium and parhelium. That the helium gas was a pure substance became clear in course of time, but physicists con-

* For a more detailed account, Grotrian, *Graphische Darstellung der Spektren*, 1928, **1** 108.

† Goudsmit and Uhlenbeck, *Physica*, 1925, **5** 266.

‡ Slater, *Nat. Acad. Sci. Proc.* 1925, **11** 732.

§ Lyman, *N*, 1922, **110** 278.

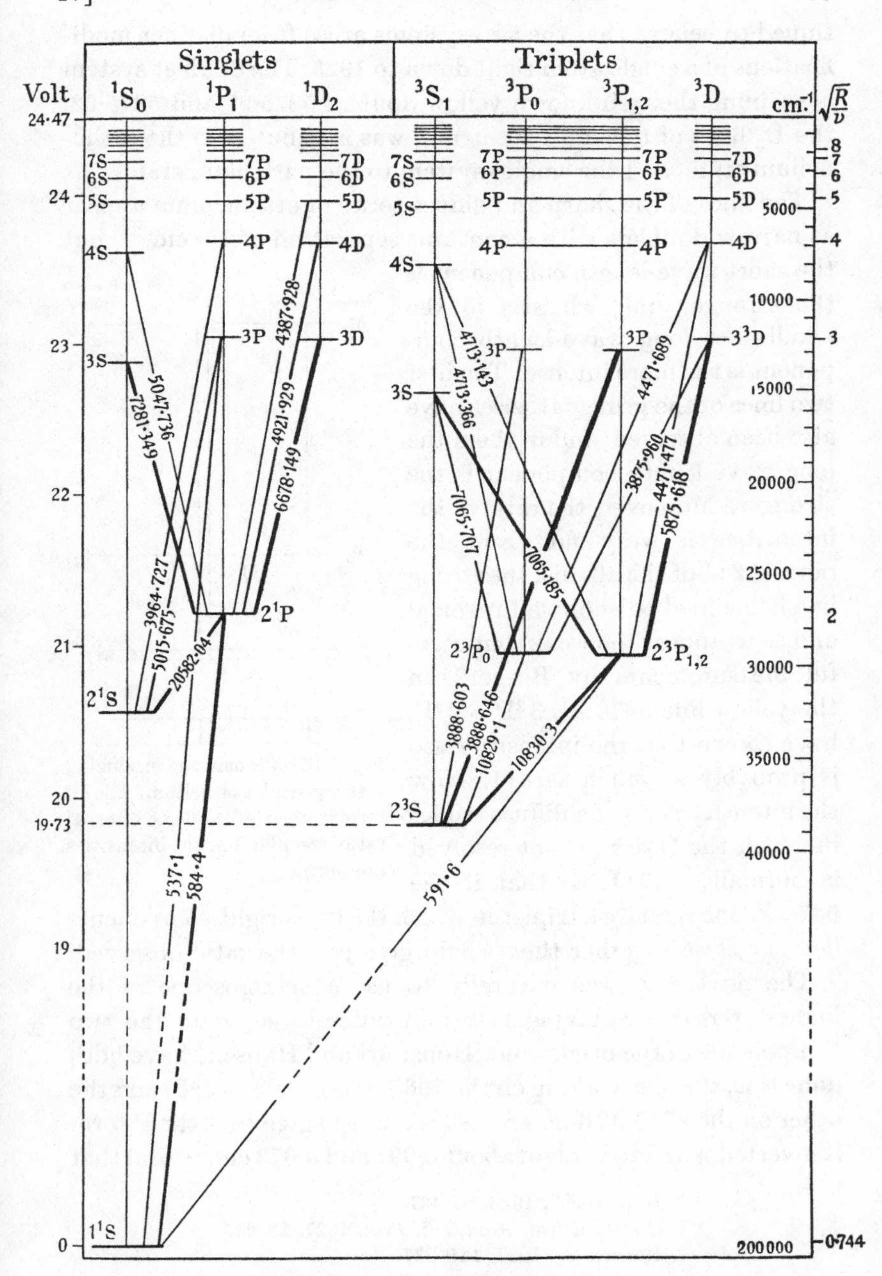

Fig. 4·14. Level diagram of helium.

tinued to believe that the two systems arose from distinct modifications of a single atom right down to 1925. The doublet system containing the well-known yellow doublet 5875·97 and 5875·62, the D_3 lines of the sun's spectrum, was attributed to the orthohelium state, and the singlet system to the parhelium state.

The lines of the sharp and diffuse series of orthohelium appear as narrow doublets with a constant separation of 1·0 cm.$^{-1}$, but the short wave-length component is the stronger line, whereas in the alkalis the long wave-length component is the more intense. The first two lines of the principal series have also been observed, and in them the long wave-length component is the stronger. Moreover, the ratio of the intensities is very far from the normal 2 : 1 of the alkalis; the strong line is inclined to show self-reversal and so to appear too weak, but careful measurements by Burger* on the yellow line 5876 A., $3\,^3D \rightarrow 2\,^3P$, have shown that the intensity ratio is probably as much as 8 : 1. Now the intensity ratio of a diffuse triplet in which the D term is not resolved is normally 5 : 3 : 1, so that if the 5876 A. line is really a triplet in which the two bright components lie very close together they would give just the ratio observed.

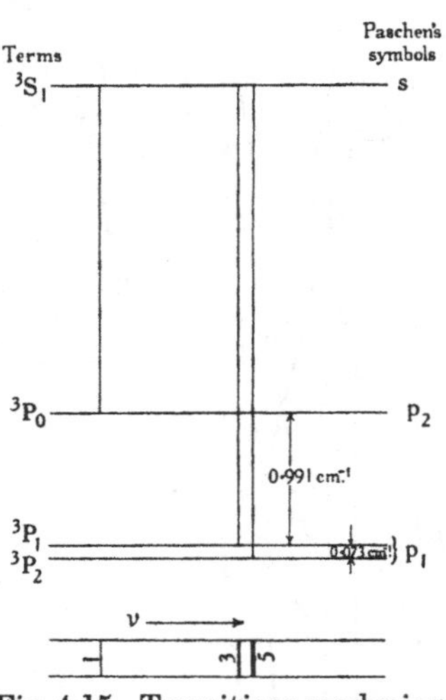

Fig. 4·15. Transitions producing a sharp triplet of helium. The ^3P term is inverted, and the interval ratio irregular, but the intensities are normal.

The next step was naturally to use a spectroscope of the highest possible resolving power in order to separate the two components of the bright line. Houston† and Hansen‡ have both done this, the one working on the 7065 A. line, $3\,^3S \rightarrow 2\,^3P$, and the other on the 4713 A. line, $4\,^3S \rightarrow 2\,^3P$. Both agree that the P term is inverted with intervals of about 0·991 and 0·073 cm.$^{-1}$, and that

* Burger, *ZP*, 1926, **38** 437.

† Houston, *Nat. Acad. Sci. Proc.* 1927, **13** 91.

‡ Hansen, *N*, 1927, **119** 237.

the intensity ratio is nearly normal (Fig. 4·15). More recently the next term of the ^3P series has been examined, in the 3888 A. line, $3\,^3P \rightarrow 2\,^3S_1$, and shown to be inverted. The intervals are $\Delta^3P_{01} = 0{\cdot}192$ and $\Delta^2P_{12} = 0{\cdot}165$ cm.$^{-1}$ *

Though this work allows the orthohelium spectrum to fit satisfactorily into its place as a product of a two-electron system, it brings out two irregularities. The $2\,^3P$ term of helium is inverted and its interval ratio is very far from the normal 2 : 1 of the alkaline earths. Whereas in calcium $\Delta P_{01} : \Delta P_{12} = 1:2$, in helium the same ratio is found to be 13 : 1. Happily Heisenberg† has deduced the interval ratio in helium from the quantum mechanics, obtaining a value of 9·5 : 1, and has shown that the normal ratio

Atomic number	Element	Ionisation potential
2	He	24·47
4	Be	9·29
12	Mg	7·61
20	Ca	6·09
38	Sr	5·67
56	Ba	5·19

Fig. 4·16. Ionisation potentials of atoms with two active electrons.

is to be expected only when the nuclear charge is large. This agrees qualitatively, if not precisely, with experiment.

With these difficulties out of the way, the term scheme of helium fits naturally into its place as the first of the alkaline earth or two-electron type. The ground term lies much deeper than in the alkaline earths, being some 20 electron volts below the $2\,^3S$ term, but an examination of the alkaline earths shows that the lower the atomic number the greater the ionisation potential (Fig. 4·16). Again there is only one inter-system combination known, a line at 591·5 A. found by Lyman and interpreted as $2\,^3P_1 \rightarrow 1\,^1S_0$, being thus homologous with the very bright 2536 A. line of mercury. It is true that in mercury many inter-system lines are known, but Heisenberg has shown that theory predicts a decrease in the intensity of inter-system

* Gibbs and Kruger, *PR*, 1931, **37** 1559.

† Heisenberg, *ZP*, 1926, **39** 499.

lines with decreasing atomic number, and in fact only one has been found in magnesium also.

The only outstanding difference is the position of the $2\,^1S_0$ and $2\,^3S_1$ terms, which lie below the P term, but this is hardly important. The selection rules do not permit a direct return from these S terms to the ground state, and in fact the lines corresponding to the transitions have not been observed.

6. Systems of higher multiplicity

In the spectra of the alkaline earths are found triplets and singlets, in the earth metals quartets and doublets, in the fourth

	Odd multiplicities							Even multiplicities							
$L \backslash J$	0	1	2	3	4	5	6	$\frac{1}{2}$	$1\frac{1}{2}$	$2\frac{1}{2}$	$3\frac{1}{2}$	$4\frac{1}{2}$	$5\frac{1}{2}$	$6\frac{1}{2}$	J / L
0	$^1S_0$						Singlets ($S=0$)	$^2S_{\frac{1}{2}}$						Doublets ($S=\frac{1}{2}$)	0
1		$^1P_1$						$^2P_{\frac{1}{2}}$	$^2P_{1\frac{1}{2}}$						1
2			$^1D_2$						$^2D_{1\frac{1}{2}}$	$^2D_{2\frac{1}{2}}$					2
3				$^1F_3$						$^2F_{2\frac{1}{2}}$	$^2F_{3\frac{1}{2}}$				3
0		$^3S_1$					Triplets ($S=1$)		$^4S_{1\frac{1}{2}}$					Quartets ($S=1\frac{1}{2}$)	0
1	$^3P_0$	$^3P_1$	$^3P_2$					$^4P_{\frac{1}{2}}$	$^4P_{1\frac{1}{2}}$	$^4P_{2\frac{1}{2}}$					1
2		$^3D_1$	$^3D_2$	$^3D_3$				$^4D_{\frac{1}{2}}$	$^4D_{1\frac{1}{2}}$	$^4D_{2\frac{1}{2}}$	$^4D_{3\frac{1}{2}}$				2
3			$^3F_2$	$^3F_3$	$^3F_4$				$^4F_{1\frac{1}{2}}$	$^4F_{2\frac{1}{2}}$	$^4F_{3\frac{1}{2}}$	$^4F_{4\frac{1}{2}}$			3
0			$^5S_2$				Quintets ($S=2$)			$^6S_{2\frac{1}{2}}$				Sextets ($S=2\frac{1}{2}$)	0
1		$^5P_1$	$^5P_2$	$^5P_3$					$^6P_{1\frac{1}{2}}$	$^6P_{2\frac{1}{2}}$	$^6P_{3\frac{1}{2}}$				1
2	$^5D_0$	$^5D_1$	$^5D_2$	$^5D_3$	$^5D_4$			$^6D_{\frac{1}{2}}$	$^6D_{1\frac{1}{2}}$	$^6D_{2\frac{1}{2}}$	$^6D_{3\frac{1}{2}}$	$^6D_{4\frac{1}{2}}$			2
3		$^5F_1$	$^5F_2$	$^5F_3$	$^5F_4$	$^5F_5$		$^6F_{\frac{1}{2}}$	$^6F_{1\frac{1}{2}}$	$^6F_{2\frac{1}{2}}$	$^6F_{3\frac{1}{2}}$	$^6F_{4\frac{1}{2}}$	$^6F_{5\frac{1}{2}}$		3
0				$^7S_3$			Septets ($S=3$)				$^8S_{3\frac{1}{2}}$			Octets ($S=3\frac{1}{2}$)	0
1			$^7P_2$	$^7P_3$	$^7P_4$					$^8P_{2\frac{1}{2}}$	$^8P_{3\frac{1}{2}}$	$^8P_{4\frac{1}{2}}$			1
2		$^7D_1$	$^7D_2$	$^7D_3$	$^7D_4$	$^7D_5$			$^8D_{1\frac{1}{2}}$	$^8D_{2\frac{1}{2}}$	$^8D_{3\frac{1}{2}}$	$^8D_{4\frac{1}{2}}$	$^8D_{5\frac{1}{2}}$		2
3	$^7F_0$	$^7F_1$	$^7F_2$	$^7F_3$	$^7F_4$	$^7F_5$	$^7F_6$	$^8F_{\frac{1}{2}}$	$^8F_{1\frac{1}{2}}$	$^8F_{2\frac{1}{2}}$	$^8F_{3\frac{1}{2}}$	$^8F_{4\frac{1}{2}}$	$^8F_{5\frac{1}{2}}$	$^8F_{6\frac{1}{2}}$	3
L / J	0	1	2	3	4	5	6	$\frac{1}{2}$	$1\frac{1}{2}$	$2\frac{1}{2}$	$3\frac{1}{2}$	$4\frac{1}{2}$	$5\frac{1}{2}$	$6\frac{1}{2}$	$L \backslash J$
$\Delta\nu$		1	2	3	4	5	6		3	5	7	9	11	13	$\Delta\nu$

Fig. 4·17. Table of spectroscopic terms.

column of the periodic system quintets and triplets; and in other elements multiplicities as high as eight occur. The singlet, doublet and triplet spectra have been satisfactorily ordered by writing the multiplicity $(2S+1)$, fixing L by the series to which a term belongs, and then allowing all those values of J which satisfy the condition

$$|(L+S)| \geqslant J \geqslant |(L-S)|.$$

Selection rules then decide which terms may be expected to combine. In analysis the determination of L and J is often difficult, but a later chapter will show how help can be obtained from the splitting of a line in a magnetic field.

If the same methods are applied to a system having a multiplicity greater than three, the number of terms found and their combinations agree precisely with those predicted by theory, the terms obtained being those set out in Fig. 4·17. This figure was first given in a slightly different form by Landé.

7. Landé's interval rule

In the sharp and principal series of calcium, the ratio of the two intervals is roughly 2 : 1. This regularity is a simple example of Landé's interval rule,* which states that in any multiplet level the separation of two terms having inner quantum number $(J-1)$ and J is proportional to J. For example:

$$^3P_2-{}^3P_1 : {}^3P_1-{}^3P_0 :: 2:1$$
$$^3D_3-{}^3D_2 : {}^3D_2-{}^3D_1 :: 3:2$$
$$^3F_4-{}^3F_3 : {}^3F_3-{}^3F_2 :: 4:3$$

Element	$\dfrac{^3P_2-{}^3P_1}{^3P_1-{}^3P_0}$	$\dfrac{^3D_3-{}^3D_2}{^3D_2-{}^3D_1}$
Beryllium	3·46	—
Magnesium	2·06	—
Calcium	2·03	1·56
Strontium	2·12	1·68
Barium	2·37	2·10
Zinc	2·05	1·62
Cadmium	2·17	1·55
Mercury	2·63	0·58

Fig. 4·18. Interval ratios of the deepest ^3P and ^3D terms of the elements of column II.

In Fig. 4·18 are given the interval ratios for the deepest P and D terms of a few triplet spectra. Deep terms will in general obey

* Landé, *ZP*, 1923, **15** 189.

the rule better than high terms, but even so, exceptions occur, particularly in elements of large atomic number.

Nevertheless the interval rule is sometimes of value in allotting J to unnamed empirical terms. Thus Fig. 4·19 shows a multiplet from the arc spectrum of iron analysed by Laporte.*

Term	$d^6\,s^2\,{}^5D$					
	J	0	1	2	3	4
$d^6\,s\,p$ ^5F	1	(30R) 26688·31 89·91	(80R) 26778·22 184·11	(20) 26962·33	—	—
	2	—	106·77 (125R) 26671·45 184·12	106·76 (100R) 26855·57 288·09	— (20) 27143·66	—
	3	—	—	164·88 (150R) 26690·69 288·07	164·90 (100R) 26978·76 415·91	— (20) 27394·67
	4	—	—	—	227·88 (200R) 26750·88 415·94	227·85 (100R) 27166·82
	5	—	—	—	—	292·29 (300R) 26874·53

Fig. 4·19. A $^5D\,^5F$ multiplet of the iron arc; the numbers in brackets are the intensities, and R means 'reversed'.

The separations of one set of terms are as 106·8 : 164·9 : 227·9 : 292·3 or as 1·8 : 2·8 : 3·9 : 5, so that presumably the values of J are 1, 2, 3, 4, 5. In tabular form this may be written

$$^5F\left\{\begin{array}{l}106{\cdot}8 \div 2 = 53{\cdot}4\\ 164{\cdot}8 \div 3 = 55{\cdot}0\\ 227{\cdot}9 \div 4 = 57{\cdot}0\\ 292{\cdot}3 \div 5 = 58{\cdot}5\end{array}\right\} \sim \text{const.} = A_F.$$

While for the combining set of terms

$$^5D\left\{\begin{array}{l}89{\cdot}9 \div 1 = \ \ 90{\cdot}0\\ 184{\cdot}1 \div 2 = \ \ 92{\cdot}1\\ 288{\cdot}1 \div 3 = \ \ 96{\cdot}0\\ 415{\cdot}9 \div 4 = 104{\cdot}0\end{array}\right\} \sim \text{const.} = A_D.$$

A glance at Landé's scheme of terms shows that the J values 1–5 and 0–4 are those predicted in the ^5F and ^5D terms.

* Laporte, *ZP*, 1924, **23** 138.

So simple a rule should clearly be explicable in terms of the vector model, and in fact the rule can be obtained provided two assumptions are made; these assumptions must be left here as if made *ad hoc*, but they arise quite logically in the theory of the wave mechanics. First assume that the energy of interaction of the orbital and spin vectors, **L** and **S**, is proportional to the cosine of the angle between them; then the energy of any level of a multiplet term will be

$$E/ch = \nu_G + ALS\cos(\mathbf{LS}), \qquad \text{......(4·1)}$$

where ν_G and A are constants whose significance will appear shortly; the product ch is introduced into the left-hand side so that these constants may be measured in cm.^{-1} As is well known in macroscopic physics $LS\cos(\mathbf{LS}) = \frac{1}{2}\{J^2 - L^2 - S^2\}$, but in atomic physics the wave mechanics shows that this must be refined to

$$LS\cos(\mathbf{LS}) = \tfrac{1}{2}\{J(J+1) - L(L+1) - S(S+1)\}, \qquad \text{...(4·2)}$$

so that the energy of the multiplet term is

$$E/ch = \nu_G + \Gamma, \qquad \text{......(4·3)}$$

where $$\Gamma = \tfrac{1}{2}A\{J(J+1) - L(L+1) - S(S+1)\}. \qquad \text{......(4·4)}$$

The numerical values obtained from this formula are given in Fig. 4·20. The separation of two components of a multiplet term, determined by J and $(J-1)$, will be given by

$$\begin{aligned}\Delta E/ch &= \tfrac{1}{2}A\{J(J+1) - (J-1)J\} \\ &= AJ, \end{aligned} \qquad \text{......(4·5)}$$

which is precisely Landé's empirical result. The constant A is thus Landé's interval quotient.

Though for the present purpose ν_G may be regarded simply as an arbitrary constant, for clarity and for future work it will be convenient to show that ν_G is the centroid of the term.

In the ^{2}P term, for example, it follows from the above equation that A is determined by

$$^2\text{P}_{1\frac{1}{2}} - {}^2\text{P}_{\frac{1}{2}} = \Delta\nu_P = \tfrac{3}{2}A.$$

And then applying equation (4·1) we obtain

$$^2\text{P}_{1\frac{1}{2}} = \nu_G + \tfrac{1}{3}\Delta\nu_P$$

and $$^2\text{P}_{\frac{1}{2}} = \nu_G - \tfrac{2}{3}\Delta\nu_P.$$

$^2P_{1\frac{1}{2}}$ stands here for the height of the level above some arbitrary zero, and not as a term value measured down from the limit.

It appears then that ν_G lies $\frac{1}{3}\Delta\nu_P$ below the higher level and divides the distance between the two lines in the ratio of 1 : 2. But this is exactly where the centroid of the two terms falls if to each term is assigned a statistical weight of $(2J+1)$, for the weights of the two ^2P terms will be 4 and 2 respectively. $(2J+1)$ is the number of Zeeman components into which a term splits in a magnetic field, and the assumption fits well both with Burger and Dorgelo's intensity rule, and with Pauli's exclusion principle.

Term \ J	0	1	2	3	4	5	6		½	1½	2½	3½	4½	5½	6½		J / Term
								Singlet								Doublet	
S	0								0								S
P		0							−1	½							P
D			0							−1½	1						D
F				0							−2	1½					F
								Triplet								Quartet	
S		0								0							S
P	−2	−1	1						−2½	−1	1½						P
D		−3	−1	2					−4½	−3	−½	3					D
F			−4	−1	3					−6	−3½	0	4½				F
								Quintet								Sextet	
S			0								0						S
P		−3	−1	2						−3½	−1	2½					P
D	−6	−5	−3	0	4				−7	−5½	−3	½	5				D
F		−8	−6	−3	1	6			−10	−8½	−6	−2½	2	7½			F
								Septet								Octet	
S				0								0					S
P			−4	−1	3						−4½	−1	3½				P
D		−8	−6	−3	1	6				−9	−6½	−3	1½	7			D
F	−12	−11	−9	−6	−2	3	9		−13½	−12	−9½	−6	−1½	4	10½		F
ΔΓ/A		1	2	3	4	5	6			1½	2½	3½	4½	5½	6½		ΔΓ/A

Fig. 4·20. The displacements of spectroscopic terms.

Since the point is important, another example may perhaps be cited; consider the ^3P term

$$^3P_2 = \nu_G + A,$$
$$^3P_1 = \nu_G - A,$$
$$^3P_0 = \nu_G - 2A.$$

And so $$^3P_2 - {}^3P_0 = \Delta\nu_P = 3A,$$

or $$A = \tfrac{1}{3}\Delta\nu_P,$$

and the centroid lines $\frac{1}{3}\Delta\nu_P$ below the highest term.

To find ν_G more directly write the wave-number of a typical term as $(\nu_G + \Gamma)$ and apply the condition that for all terms of the multiplet

$$\Sigma(2J+1)\Gamma = 0.$$

In the ^3P term, for example, this gives

$$5\Gamma_2 + 3\Gamma_1 + \Gamma_0 = 0.$$

And if the term has the ideal separations of the interval rule

$$(\Gamma_2 - \Gamma_1) = 2(\Gamma_1 - \Gamma_0) = \tfrac{2}{3}\Delta\nu_P,$$

where $\Delta\nu_P = {}^3P_2 - {}^3P_0$ is the extreme interval of the multiplet. From this it follows that

$$\Gamma_2 = \tfrac{1}{3}\Delta\nu_P, \quad \Gamma_1 = -\tfrac{1}{3}\Delta\nu_P, \quad \Gamma_0 = -\tfrac{2}{3}\Delta\nu_P.$$

If a term is inverted so that the term with the greatest J lies deepest, the values of Γ will be reversed in sign but unaltered in magnitude; while if the separations are not in the ideal ratio of Landé's rule, then the actual separations should be used in the calculation.

It does not seem possible to give a simple proof that the two methods of arriving at the centroid of a term will always give the same result, though a proof for a particular value of S is simple enough; but a proof is perhaps unnecessary where a check is so easily applied.

BIBLIOGRAPHY

The most complete account of the alkaline earths, as of the alkalis, is Grotrian, *Graphische Darstellung der Spektren von Atomen mit ein, zwei und drei Valenzelectronen*, 1928.

CHAPTER V

ABSORPTION SPECTRA AND RELATED PHENOMENA

1. Absorption spectra

More than a century ago Fraunhofer* named a number of dark lines which Newton had noticed in the spectrum of the sun. In the laboratory these dark lines may be demonstrated; if a sodium flame is placed in front of a spectrometer, two bright lines appear; but if a gas-filled electric bulb or an arc is placed behind the flame and focused through it on to the slit, the sodium lines appear in the same place, dark on a bright background. Thus the dark lines are clearly due to absorption of light from the hot filament in the relatively cold flame; and in fact the Fraunhofer lines have long been recognised as coinciding with the emission lines of such elements as sodium, iron and calcium.

Since any line appearing in absorption must show a transition from the state in which the atom happens to be to some higher state, one might expect all absorption lines to arise in the lowest level of the spectrum, for this will be the normal or 'ground' state of the atom. And in fact in the spectra of the alkalis and alkaline earths the principal series appears in absorption, though the sharp and diffuse series do not; Wood and Fortrat,† for example, used a column of non-luminous sodium vapour 2·8 metres long with a spectrograph of such high dispersion that 1 A. appeared as 3·5 mm. on the plate, and distinguished 58 lines of the principal series $m\,{}^2P \leftarrow 1\,{}^2S_{\frac{1}{2}}$.

Needless to say this technique can supply valuable evidence when the ground term happens to be in doubt; in aluminium, for example, the lowest known term is ^2P, but one might reasonably suppose that a lower ^2S term exists lying so low that all combinations are in the ultra-violet; Grotrian,‡ however, has shown

* Fraunhofer, *Gibbert's Annalen*, 1817, **56** 264. Cf. Kayser, *Handbuch d. Spektroskopie*, 1900, **1** 7, 9.

† Wood, R. W. and Fortrat, *AJ*, 1916, **43** 73.

‡ Grotrian, *ZP*, 1922, **12** 218.

that the sharp and diffuse series appear in absorption, though the principal series does not, and this is conclusive; the ground term is ^2P. On the other hand, in a thorough analysis of the visible spectrum of neon, the lowest terms which Paschen discovered were four which he labelled s_5, s_4, s_3 and s_2; yet lines arising from these terms appear in absorption only when the neon is previously excited;* and in fact some years later Hertz found in the far ultra-violet two lines, which show that the ^1S ground term lies so low that it produces no line of wave-length longer than 743 A.†

Under certain conditions however lines, which do not arise in the ground term, appear in absorption; the neon lines arising in the s terms have already been noticed, while the Balmer series appears sharply reversed in the spectra of certain stars. Moreover, in the laboratory lines often appear 'self-reversed', that is to say a narrow black line appears down the centre of an emission line, an appearance which is interpreted to mean that the hot vapour in the centre of the tube emits a broad line due to a large Doppler effect, while the cooler vapour near the walls absorbs a narrower band.

As the reversal of the Balmer series presumably means that in the outer layers of the star a proportion of the hydrogen atoms are in the excited state with $n=2$, many have wished to reproduce the phenomenon under conditions which can be more conveniently examined. The first requisite is to obtain a large number of atoms in the excited state, an end which may be achieved by raising the temperature or by using cold vapour and exciting it with electrons or radiation. This is not the place to give an exhaustive account of this work, but only to illustrate the mechanism, so one example of each method must suffice.‡

The changes produced by a rise in temperature appear most clearly when there are only two low levels and these lie fairly close together as they do in the ^2P ground term of the earth metals, indium and thallium. $^2P_{\frac{1}{2}}$ is the lower of these levels, while $^2P_{1\frac{1}{2}}$

* Meissner, *AP*, 1925, **76** 124.

† Hertz, *ZP*, 1925, **32** 933.

‡ Andrade, *Structure of the atom*, 1927, 304f.; Pringsheim, *Fluorescenz und Phosphorescenz*, 1928, 26f.

is metastable since the L selection rule does not allow the two to combine; if energy is gained in an inelastic collision, however, the selection rules are of no account, so that the random collisions of a gas keep a certain number of atoms permanently in the higher state; this number should increase with increase of temperature, but should be less the greater the height of the $^2P_{1\frac{1}{2}}$ level above the ground state. Experimenting, Grotrian* was able to divide the absorption spectrum into two sets of lines; at 400° C. thallium vapour absorbs the lines 3776 and 2768 A., but when the temperature is raised to 800° C. the line 5350 A. appears together with the doublet 3519 and 3529 A.; and analysis shows that the two lines appearing at 400° C. arise as $2\,^2S_{\frac{1}{2}} \leftarrow 2\,^2P_{\frac{1}{2}}$ and $3\,^2D_{1\frac{1}{2}} \leftarrow 2\,^2P_{\frac{1}{2}}$, while the lines appearing only when the temperature is raised are $2\,^2S_{\frac{1}{2}} \leftarrow 2\,^2P_{1\frac{1}{2}}$ and $3\,^2D_{2\frac{1}{2},1\frac{1}{2}} \leftarrow 2\,^2P_{1\frac{1}{2}}$. In indium the corresponding temperatures are 650° and 800° C., so that the temperature difference increases from 150° to 400° C. as the ^2P interval increases from 112 cm.$^{-1}$ in indium to 7793 cm.$^{-1}$ in thallium.

In these experiments both the term intervals and the temperature difference are small, but there is no reason to doubt that the mechanism is the same as that which causes the absorption of the Balmer series in stellar spectra. Indeed the Boltzmann distribution law allows a general account of the phenomena, for it states that if an atom can exist in two states having energies E_1 and E_2, the number of atoms in the states will be in the ratio

$$a_1 e^{-E_1/k\theta} : a_2 e^{-E_2/k\theta},$$

where a_1, a_2 are the statistical weights, θ the absolute temperature, and k the Boltzmann constant. Now the statistical weight of an atomic level is simply $(2J+1)$, so that in Grotrian's experiment the ratio of the number of atoms in the $^2P_{1\frac{1}{2}}$ state to the number in $^2P_{\frac{1}{2}}$ is $3e^{(E_{\frac{1}{2}}-E_{1\frac{1}{2}})/k\theta}$.

Atoms may be raised to a higher level by the collisions of a discharge tube, as well as by a rise of temperature. To illustrate this Metcalfe and Venkatesachar† enclosed some mercury vapour in a long tube, and passed through it a small current

* Grotrian, *ZP*, 1922, **12** 218.

† Metcalfe and Venkatesachar, *PRS*, 1921, **100** 149.

adjusted to make the vapour faintly luminous. Lines arising in any of the 2 ^{3}P levels are strongly absorbed by this tube, but those arising in the higher 2 ^{1}P level are not absorbed at all (Fig. 5·1). And this we might expect, for with a low potential gradient an atom will normally make several collisions before it has acquired the 4·9 volts needed to lift it to one of the ^{3}P levels, but as all

I. *Lines strongly absorbed*

5461	$2\,^3S_1 \leftarrow 2\,^3P_2$
3342	$3\,^3S_1 \leftarrow 2\,^3P_2$
3663, 3655, 3650	$3\,^3D_{1,2,3} \leftarrow 2\,^3P_2$
4359	$2\,^3S_1 \leftarrow 2\,^3P_1$
3132, 3126	$3\,^3D_{1,2} \leftarrow 2\,^3P_1$
4047	$2\,^3S_1 \leftarrow 2\,^3P_0$
2967	$3\,^3D_1 \leftarrow 2\,^3P_0$

II. *Lines not absorbed*

4916	$3\,^1S_0 \rightarrow 2\,^1P_1$
5791	$3\,^1D_2 \rightarrow 2\,^1P_1$
4347	$4\,^1D_2 \rightarrow 2\,^1P_1$
5770	$3\,^3D_2 \rightarrow 2\,^1P_1$
4339	$4\,^3D_2 \rightarrow 2\,^1P_1$
4078[a]	$2\,^1S_0 \rightarrow 2\,^3P_1$

Fig. 5·1. Lines absorbed by cold mercury vapour made faintly luminous by an electric current.

[a] The absence of the 4078 A. line in absorption is anomalous. Metcalfe and Venkatesachar, *PRS*, 1921, **100** 149.

these collisions are elastic they do not absorb any of the energy the atom has acquired; when an atom already carries more than 4·9 volts, however, it is likely to lose a large part of this whenever it collides, so that the chance of its acquiring the 6·7 volts needed to lift it to the ^{1}P level is very small.

Using a similar technique Pflüger showed that electrically excited hydrogen absorbs the Balmer series.*

Thus atoms which have been excited by an electron discharge do absorb the lines which the quantum theory suggests. No one seems yet to have produced an absorption line when the preliminary excitation is optical, but Füchtbauer has been able to show that only when mercury vapour is excited by the 2537 A. line can it absorb lines whose lower level is ^{3}P. The illumination must be intense if sufficient atoms are to be in the excited state,

* Pflüger, *AP*, 1907, **24** 515. Cf. Ladenburg, *Verh. Phys. Ges.* 1910, **12** 549.

for after a very short period each atom will radiate and return to the ground level; accordingly Fuchtbauer* passed a discharge through the space between two coaxial tubes, and placed a cylinder containing mercury vapour along their axis. All the tubes were of quartz, so that they would pass the 2537 A. line, but not 1849 A. or any line of shorter wave-length. When the pressure was suitably adjusted, the vapour in the central cylinder emitted all the arc lines as soon as the outer arc was struck; but when a thin glass tube was inserted between the discharge tube and the central cylinder, the emission ceased. The explanation is simple. The atoms in the central cylinder are in the normal ^1S state and can absorb only lines which originate in this state; of these the line of longest wave-length is 2537 A. and the next 1849 A., lines which lift the atom to the $^3P_1$ and $^1P_1$ states; thus when there is no glass between the lamp and the cold vapour, the 2537 A. line lifts the atoms into the $^3P_1$ state, and they can then absorb any lines which start in this state. A brief examination of the level scheme (Fig. 4·7) shows that by absorbing one more line and emitting another, these atoms can reach any level whatever, so that a complete emission spectrum may be expected.

This experiment leaves no doubt that atoms previously excited by light do absorb from the state which they have reached, though the concentration in the excited state may not be sufficient to produce an absorption line on a continuous background.

2. Ionisation and resonance potentials

The energy required to ionise an atom or to raise it to an excited state can be calculated from the spectrum, or it can be obtained quite independently by exciting the atom with electrons which have fallen through a known potential difference.

To illustrate the spectroscopic method, consider the sodium lines 5890 and 3302·3 A., which analysis reveals as the first two members of the principal series; they differ in frequency by 13300 cm.$^{-1}$, so that the series should have a limit 24643 cm.$^{-1}$ above the $2\,^2P_{1\frac{1}{2}}$ term; this value can be obtained by the use of Rydberg's law, or it can be read off more simply from his tabula-

* Fuchtbauer, *PZ*, 1920, **21** 635.

tion of the values of $R/(n+a)^2$; it does not agree precisely with the accepted value of 24476 cm.$^{-1}$, but only because the series does not follow Rydberg's law precisely; in fixing the accepted value, the higher series terms have been considered. As the ground state of sodium is $1\,^2S_{\frac{1}{2}}$, the energy required to raise the atom to the $2\,^2P_{1\frac{1}{2}}$ state is $10^8/5890$ or 16973 cm.$^{-1}$, and the ionisation potential is this added to 24476 or 41449 cm.$^{-1}$.

In order to compare these spectroscopic values with those obtained by exciting the atom with high-speed electrons, consider an electron which has fallen through 1 volt; its energy is

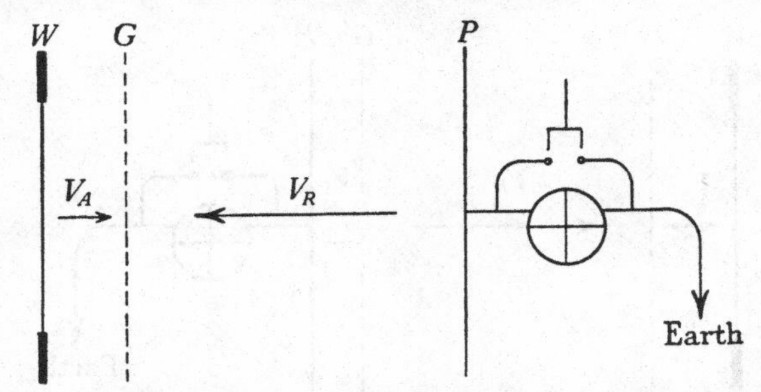

Fig. 5·2. Set-up used by Lenard to measure excitation potentials.

$\frac{4{\cdot}77\,.\,10^{-10}}{300}$ ergs, so that it can excite a line of wave-number ν, where ν is given by $\quad 1{\cdot}59\,.\,10^{-12} = ch\nu,$

c being the velocity of light and h Planck's constant. When the usual values are assigned to the constants, this gives

$$\nu = \frac{1{\cdot}59\,.\,10^{-12}}{3\,.\,10^{10}\,.\,6{\cdot}55\,.\,10^{-27}} = 8106\,\text{cm.}^{-1}$$

Thus an electron which has fallen through 1 volt can excite a line of wave-length 12336 A. The ionisation potential of sodium is consequently $41449/8106 = 5{\cdot}11$ volts, and the resonance potential of the $2\,^2P_{1\frac{1}{2}}$ state $12336/5890 = 2{\cdot}10$ volts.

The electrical methods have all been inspired by the work of Lenard,* who in 1902 accelerated the electrons emitted by a hot wire W (Fig. 5·2) with a potential V_A applied between the wire and a gauze G; these electrons are prevented from reaching the

* Lenard, *AP*, 1902, **8** 149.

plate P of the electroscope by a retarding potential V_R, which is greater than V_A. If then the electroscope receives any charge, it must depend on one of two mechanisms; either the electrons must excite the atoms present to emit light and this falling on the plate causes it to emit electrons; or the high-speed electrons must ionise the atoms near the gauze and the positive ions produced then fall on to the plate.

This simple technique however does not distinguish between resonance and ionisation potentials, and accordingly Davis and Goucher* introduced a second gauze G' (Fig. 5·3), which they

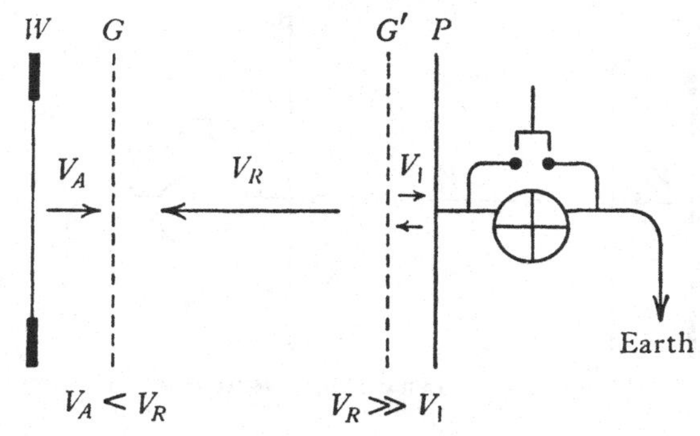

Fig. 5·3. Set-up used by Davis and Goucher to distinguish ionisation from resonance potentials.

kept at a potential slightly greater or slightly less than that of the plate. Whether this small potential is positive or negative makes little difference to any positive ions which are liberated, for having fallen through most of V_R by the time they reach G', they have enough energy to overcome a small opposing potential; but it suffices to control the direction of any photoelectric current; for if G' is positive, the photoelectrons escape from P as before and the electroscope receives a positive charge; while if G' is negative, no photoelectrons liberated from P can escape, while any electrons liberated on the side of G' near to P pass to the electroscope giving it a negative charge. In accord with this description, Davis and Goucher obtained for mercury the curves shown in Fig. 5·4; in the upper curve G' is positive to P, while in

* Davis and Goucher, *PR*, 1917, **10** 101.

the lower curve it is negative. That the current is zero so long as V_A is less than 4·9 volts suggests that this is the lowest resonance potential of the mercury atom, while the sudden rise in the current when the potential reaches 6·7 volts suggests a further resonance potential at this value; the upward turn in the lower curve shows that 10·4 volts is the ionisation potential.

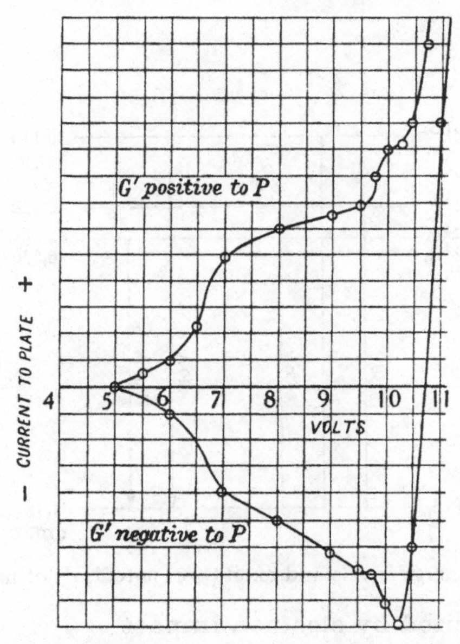

Fig. 5·4. Resonance and ionisation potentials of mercury vapour. After Davis and Goucher, *PR*, 1917, **10** 106.

These measurements receive a simple explanation in terms of spectroscopic theory; the normal state of the mercury atom is $^1S_0$, while next above this lies the $2\,^3P$ term, the line resulting from the transition $2\,^3P_1 \rightarrow 1\,^1S_0$ being 2537 A., a figure which enables us to calculate the potential required as 4·86 volts (Fig. 5·5). Of the levels which can combine with the ground state, the next above $2\,^3P_1$ is $2\,^1P_1$, and as an atom leaving this emits the 1849 A. line, the potential required to excite it is 6·67 volts. The ionisation potential obtained spectroscopically is 10·39 volts. All these values agree admirably with the measurements of Davis and Goucher.

The experimental method described above is not the most sensitive now available, but it is simple and serves to illustrate the principles on which all electrical methods depend.

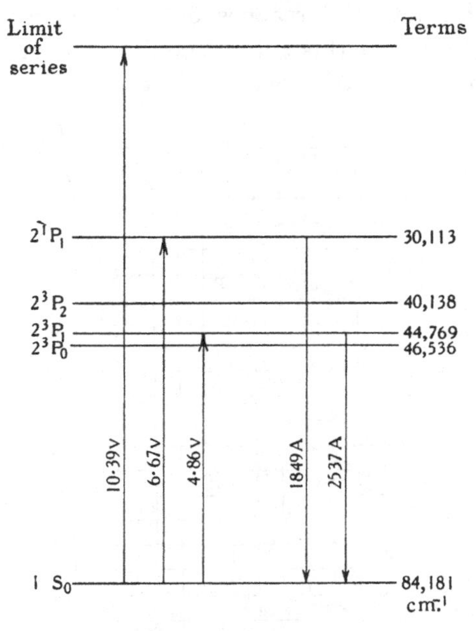

Fig. 5·5. Energy levels and excitation potentials of mercury.

3. Spectra excited by electron impact

Interesting confirmation of atomic energy levels can be obtained, if the spectrum is excited by electrons which have fallen through a known voltage. Franck and Hertz* first showed that when low-pressure mercury vapour is excited in this way, it emits no radiation at all until the potential rises to 4·9 volts, and that then only one line, 2537 A., appears. This accords well with theory, for a line of this wave-length implies an energy change of 4·86 volts, and the $2\,^3P_1$ level, in which this line arises, is the lowest from which a transition back to the ground term is permitted. Similar one-line spectra have been produced in zinc, magnesium and other metallic vapours.†

* Franck and Hertz, *PZ*, 1919, **20** 132.

† Fridrichson, *ZP*, 1930, **64** 43, gives references.

Whatever the details of the apparatus used, the distance between the filament and grid must be small, so that very few atoms shall be struck before the electrons have acquired their full velocity; on the other hand the space between the grid and plate must be large, so that sufficient atoms may be excited. Newman* (Fig. 5·6), for example, used three tungsten wires F as a source of electrons, and placed in front a nickel gauze G, arranged so that the distance between the two was not more than 1·5 mm. The gauze, which supplied the accelerating potential, was electrically connected to the nickel cylinder A, so that the space between G and A was a region of constant potential. The whole tube was enclosed in an electric furnace, whose interior was maintained at 350° C.; at this temperature the vapour pressure of the sodium

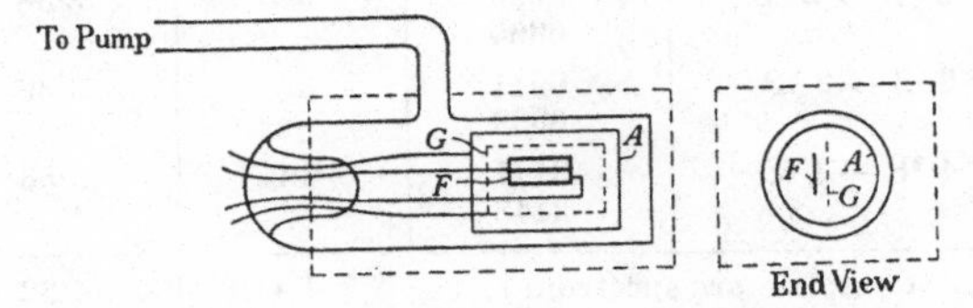

Fig. 5·6. Newman's discharge tube. *PM*, 1925, **50** 170.

enclosed is less than $\frac{1}{10}$ mm. of mercury. Raising the voltage step by step, Newman showed how in sodium new lines appear at definite potentials. Unfortunately the photographic plates have been destroyed, so that only a verbal description of the spectrograms, published in the *Philosophical Magazine*, can be given here. In the first the voltage is 2·2 and only the doublet 5896–90 A. appears;† at 3·18 and 3·6 volts the doublets 11404–382 and 8195–83 A. respectively should appear, but both are beyond the range of the plate, so that at 3·74 volts the only lines on the plate should be 5896–90 and 3303–02 A., and in fact the second spectrogram taken at 4·0 volts shows these lines and no others. The lines 6161–54 A. should be the next to appear, and the third spectrogram taken at 4·4 volts contains them, while the further doublet 5688–83 A. appears at 4·6 volts. With the exposures employed no further lines appear before the ionisation voltage of 5·2 is

* Newman, *PM*, 1925, **50** 165.

† For the level diagram of sodium, see Fig. 3·7.

reached, but the results suffice to show that the lower levels at least can be excited one by one. Equally satisfactory agreement was obtained with potassium and rubidium,* the measurements on the former being tabulated in Fig. 5·7.

Transition	Wave-length (A.)	Excitation voltage	
		Observed	Calculated
$2\,^2P \rightarrow 1\,^2S$	7665 7699	1·9	1·61
$2\,^2S \rightarrow 2\,^2P$	12434 12523	—	2·59
$3\,^2D \rightarrow 2\,^2P$	11690 11772	—	2·66
$3\,^2P \rightarrow 1\,^2S$	4044 4047	3·3	3·05
$3\,^2D \rightarrow 2\,^2P$	6936 6965	3·7	3·38
$3\,^2S \rightarrow 2\,^2P$	6911 6939	3·7	3·39
$4\,^2P \rightarrow 1\,^2S$	3447 3448	3·9	3·58
Complete arc spectrum		4·4	4·32

Fig. 5·7. Line by line excitation of potassium.

4. Spectra excited by monochromatic light†

When sodium vapour at a low pressure is illuminated with the yellow D lines, it re-emits the D lines in all directions; this re-emitted light has been called 'resonance radiation', since on the wave theory it is explained as due to vibrators within the atom which respond to waves of their own period.

Pure resonance radiation can be obtained only if atoms leaving the excited state have no alternative but to return to the ground state whence they came. As an excited atom may lose energy by collision or radiation, this condition has two consequences; first, the vapour pressure must be so low that the time between successive impacts is small compared with the life of an excited atom; and second, if the excited level is not the first above the ground

* Newman, *PM*, 1925, **50** 796, 1276.

† Andrade, *The structure of the atom*, 1927, 358–60; Pringsheim, *Fluorescenz und Phosphorescenz*, 1928, 20–24.

level, the selection rules must not allow the atom to pass from the excited to the intermediate level.

Wood and Mohler* have shown that when sodium vapour at 200° C. is illuminated with one D line, only that line is emitted; but if some hydrogen is left in the bulb, or if the temperature is raised to 300° C. so that the vapour pressure rises from 3.10^{-3} to $2{\cdot}5.10^{-2}$ mm., both lines appear showing that a number of atoms suffer an inelastic collision before they radiate. This illustrates

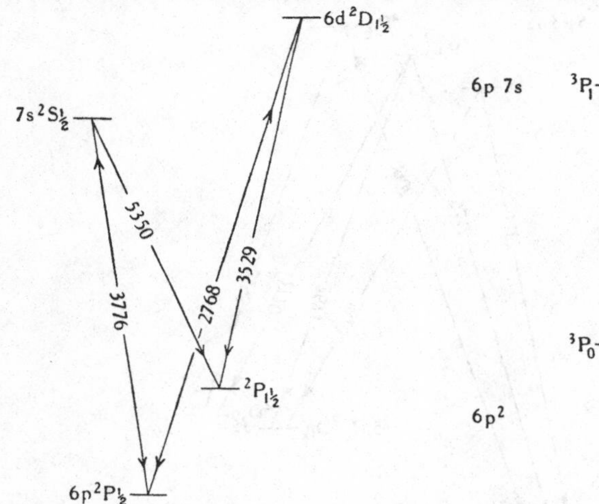

Fig. 5·8. Fluorescence of thallium vapour; the upward pointing arrows show the lines absorbed, the downward the lines emitted.

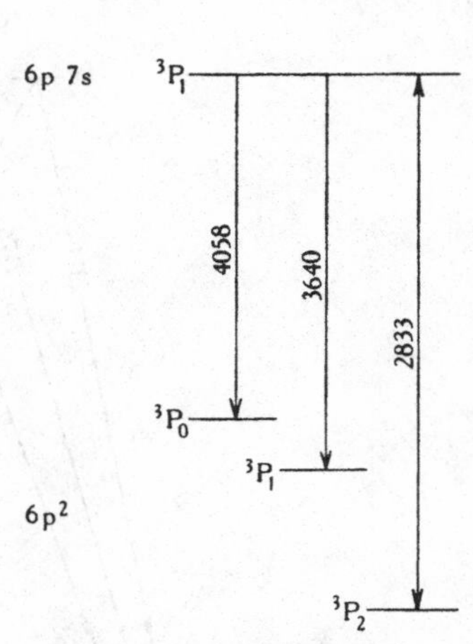

Fig. 5·9. Fluorescence of lead vapour.

the first condition; in illustration of the second consider the 1849 A. line of mercury; it should be a resonance line like 2537 A., for it arises as $2\,^1P \rightarrow 1\,^1S$, and the only term between the two is $2\,^3P$ with which $2\,^1P$ cannot combine. Rump† has confirmed this prediction, but the work is not easy, for the 1849 A. line lies in a region in which oxygen absorbs, so that the apparatus must be filled with carbon dioxide.

In the two examples cited, the ground term has been single, for resonance in the strict sense is seldom observed when the ground

* Wood, R. W. and Mohler, *PR*, 1918, **11** 70.

† Rump, *ZP*, 1925, **31** 901.

term is multiple; instead the vapour is fluorescent. Thallium vapour, for example, whose normal state is $2\,^2P_{\frac{1}{2}}$, may be raised to the $2\,^2S_{\frac{1}{2}}$ or $3\,^2D_{1\frac{1}{2}}$ state by irradiating it with the 3776 or 2768 A. line, and this line will be re-emitted (Fig. 5·8), but with it will appear the 5350 or 3529 A. line, for some atoms will pass to the $2\,^2P_{1\frac{1}{2}}$ state.* Closely analogous results are obtained in lead and antimony; if lead vapour is irradiated with the line 2833 A., it is re-emitted accompanied by two other lines 3640 and 4058 A.

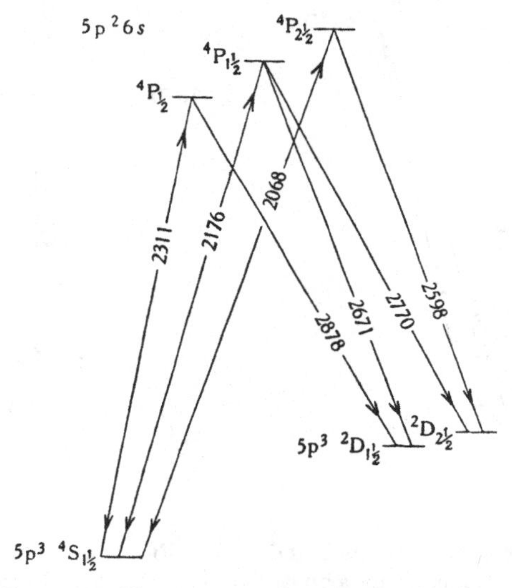

Fig. 5·10. Fluorescence of antimony vapour.

(Fig. 5·9); while the observations on antimony are adequately summarised in Fig. 5·10. An upward pointing arrow shows that the line was exciting and a downward that it was being emitted.

Fluorescence may be excited as well by a line which raises the atom to a high level, as by one which raises it only to a low. Lord Rayleigh† showed this when he irradiated sodium vapour with the 3304 and 3302 A. lines, thus raising the atoms direct to the $3\,^2P$ state, and found that this causes the emission of the two D lines. As the direct transition from $3\,^2P$ to $2\,^2P$ is forbidden, there

* Terenin, *ZP*, 1925, **31** 26; 1926, **37** 98. Terms of Sb from Charola, *PZ*, 1930, **31** 457.

† Strutt, *N*, 1915, **95** 285; *PRS*, 1919, **96** 272.

seems little doubt that the atoms either lose energy in a collision or pass from one to the other by way of the 2 ^{2}S or 3 ^{2}D state, though all intermediate transitions produce lines of such long wave-length that Lord Rayleigh did not observe them. Either mechanism explains the fact that when the vapour is irradiated with a zinc line, which happens to coincide with the 3304 A. line, both D lines appear.

BIBLIOGRAPHY

The aim of this chapter is to show that the energy levels postulated to explain emission spectra can be confirmed by other methods. No section pretends to be more than a cursory account of the subject. General works which take up these matters more thoroughly are:

Franck and Jordan. *Anregung von Quantensprüngen durch Stösse.* 1926.
Andrade. *The structure of the atom.* 1927.
Pringsheim. *Fluorescenz und Phosphorescenz.* 1928.
Jordan. "Energiestufen in Spektren." *Handbuch d. Phys.* 1929, **21** 463–92.
Arnot. *Collision processes in gases.* 1933.
Mitchell and Zemansky. *Resonance radiation and excited atoms.* 1934.

CHAPTER VI

THE ZEEMAN EFFECT

1. The normal Zeeman triplet

If a source of light, such as a discharge tube or a spark, is placed between the poles of an electromagnet, and examined with a spectroscope, one observes that as the current in the coils of the magnet is increased, the lines appear to broaden, and finally if the resolving power is sufficiently high, to split into a number of components. This effect was first observed by Zeeman* in 1896.

By chance Zeeman happened to be working with lines of a singlet series, and so observed that the single line splits into three polarised components; both splitting and polarisation Lorentz explained with the classical electromagnetic theory. A year later, however, Zeeman and other workers showed that the lines of other series split not to the simple triplet, but to a group of four or more components; and as the classical theory was unable to account for this, the two types were labelled the 'normal' and 'anomalous' Zeeman effects.

When a spectral line splits in a magnetic field, the lines into which it splits, the so-called Zeeman components, are polarised either parallel or perpendicular to the magnetic field; the former are commonly referred to as π and the latter as σ components, σ deriving from the German 'senkrecht'. In a figure, such as Fig. 6·1, drawn to show the splitting of a line, the π components are drawn above the horizontal, and the σ components below, the lengths of the verticals being proportional to the intensities.

The Zeeman components are always symmetrical about the position of the undisplaced line, and accordingly in the normal triplet the three components arising from the line of frequency ν' have frequencies ν' and $\nu' \pm \Delta\nu'$. The undisplaced line is a π component, and the two displaced lines are σ components. Thus an observer looking along a line perpendicular to the field of force

* Zeeman, *PM*, 1897, **43** 226. A translation from *Zitt. Akad. Amsterdam*, 1896, **5** 181, 242.

sees a central component polarised parallel to the field and on either side a component polarised perpendicular to the field; while if he looks along the lines of force through a hole bored in one of the pole-pieces, the central line disappears and the two lateral

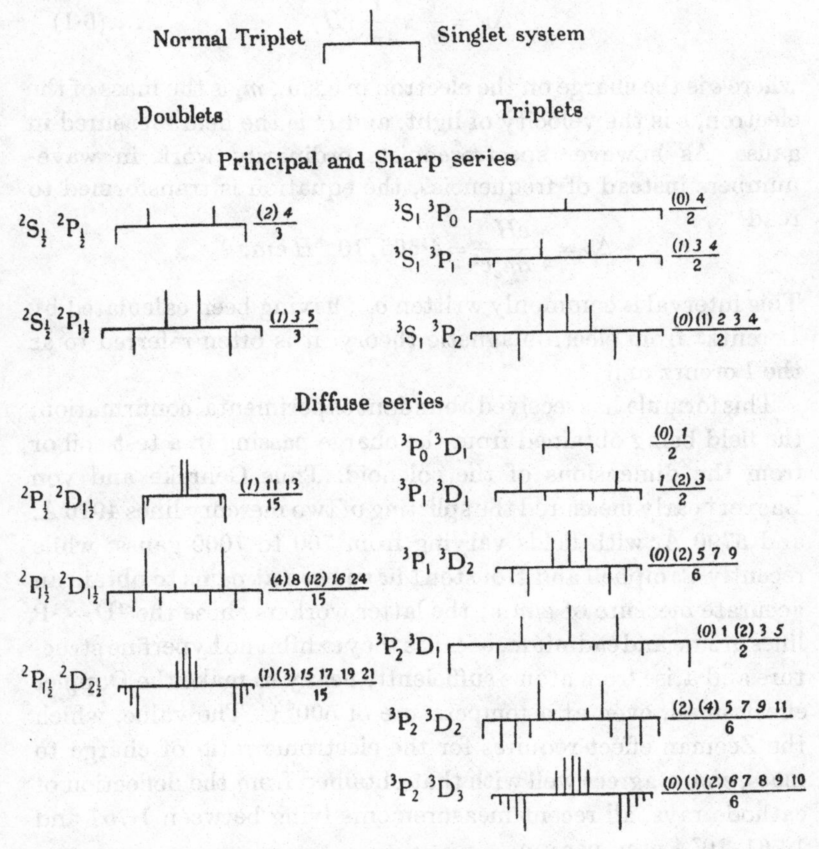

Fig. 6·1. Zeeman patterns. The π components are drawn above the line, σ components below; the heights show the intensities.

components are circularly polarised; further the latter rotate in opposite directions.

Photographs are commonly taken perpendicular to the lines of force, for they show all the components, and the mere rotation of a nicol distinguishes the π and σ polarisations; moreover, a hole bored in a pole-piece inevitably destroys the uniformity of the magnetic field.

The displacement $\Delta\nu'$ of each σ component is proportional to the strength of the magnetic field, and is given by the expression

$$\Delta\nu' = \frac{e}{m_e} \cdot \frac{1}{4\pi c} . H, \qquad \ldots\ldots(6\cdot 1)$$

where e is the charge on the electron in e.s.u., m_e is the mass of the electron, c is the velocity of light, and H is the field measured in gauss. As however spectroscopists ordinarily work in wave-numbers instead of frequencies, the equation is transformed to read

$$\Delta\nu = \frac{eH}{4\pi m_e c^2} = 4\cdot 665 . 10^{-5} H \text{ cm.}^{-1}$$

This interval is commonly written o_m; having been calculated by Lorentz* from electromagnetic theory, it is often referred to as the Lorentz unit.

This formula has received abundant experimental confirmation, the field being obtained from the charge passing in a test coil or from the dimensions of the solenoid. Thus Gehrcke and von Baeyer† early measured the splitting of two mercury lines 4916 A. and 5790 A. with fields varying from 700 to 7000 gauss; while recently Campbell and Houston‡ have been at pains to obtain an accurate measure of $e/m_e c$; the latter workers chose the $^1D \rightarrow {}^1P$ lines in zinc and cadmium, because they exhibit no hyperfine structure and arise from atoms sufficiently heavy to make the Doppler effect small, even at a temperature of 500° C. The value, which the Zeeman effect requires for the electronic ratio of charge to mass, $e/m_e c$, agrees well with that obtained from the deflection of cathode rays, all recent measurements lying between 1·757 and 1·761 . 10^7 e.m.u. per gm.

The displacement in the simple triplet is very small, being only 1·9 cm.$^{-1}$ or one-ninth of the $D_1 D_2$ interval, even when the electromagnet produces a field of 40,000 gauss; and few

* Lorentz, *K. Akad. Amsterdam*, 1897, **6** 193.

† Gehrcke and von Baeyer, *AP*, 1909, **29** 941. An earlier absolute measurement is that of Weiss and Cotton, *J. de Phys.* 1907, **6** 429.

‡ Campbell and Houston, *PR*, 1932, **39** 601.

PLATE III. MAGNETIC SPLITTING PATTERNS

1. [illegible] 6438·5 Å, both π and σ components 6438·496 [illegible].

2. [illegible] $\to$ [illegible] Cd 4678·1 Å; normal triplet (0)1/1, π components above, σ below.

3. $^1P_1 \to {}^1S_0$ Ne 6266·5 Å, [illegible] in left; $^3P_1 \to {}^3P_2$ Ne 6598·95 Å, [illegible] line on right; the line in the absence of magnetic field below, π and σ components together above.

4. $^3S_1 \to {}^3P_0$ Zn 4680·1 Å; pattern (0) 2/2.

5. $^3S_1 \to {}^3P_1$ Zn 4722·2 Å; pattern (1) 3 4/2.

6. $^3S_1 \to {}^3P_2$ Zn 4810·5 Å; pattern (0)(1) 1 2 3 4/2.

7. (^{4}[illegible]) $\to$ (^4S) [illegible] O [illegible] 4368·3 Å, pattern (0)(1)(2) 4 5 6 7 8/3, π components above, σ below.

8. $^4P_{5/2} \to {}^4S_{3/2}$ Cr 4275·0 Å; pattern (1)(3)(5) 21 23 25 27 29/15. The components separated by an interval of [illegible] are not perfectly resolved, but the separations of the strongest σ components can be accurately measured.

9. $^6P_{7/2} \to {}^6S_{5/2}$ Cr 5208·4 Å; pattern (0)(1)(2)(3) 4 5 6 7 8 9 10/[illegible]. The weakest components of 10/[illegible] do not appear in the reproduction.

10. $^3D_3 \to {}^3P_2$ [illegible] 4579·6 Å; pattern (4) 8 (2)(7) 8 9 10 [illegible]; components above, π below.

11. $^3D_1 \to {}^3P_1$ Zn 3282·3 Å; pattern (0) 1/2; lines in absence of field above, π and σ components together below. In the lower figure the forbidden transition $^3D_2 \to {}^3P_1$ with pattern (0) 2/6 appears, while in the upper figure it is absent.

12. $5p$ $^3F \to 6s$ ^3D, Pb 4063·6 Å. The observed pattern (0)(1)(2)(3) [illegible] [illegible] 1·23 1·85 [illegible] have evaluated by g-factors of 0·89 and 1·22. Both these values are in agreement with the vector coupling rule (§5).

13. $^3D_2 \to {}^3P_1$ Zn 3345·6 Å and $^3D_3 \to {}^3P_2$ [illegible] 3403·0 Å, on right; both patterns 3/2; σ and π components together. Both photographs were taken in a field of 40,000 gauss, making a [illegible] of the Paschen-Back effect is more obvious in Zn than in Cd, because the ^3D interval factor is 1/3 of that in Zn and 6/3 in Cd.

14. $^3D_3 \to {}^3P_2$ Zn 3282·3 Å, on left; $^3D_2 \to {}^3P_1$ Cd 3403·6 Å, on right. In each figure the intense pattern is (0)(2) 4 7 9/6 from $^3D_3 \to {}^3P_2$, and the faint (1)(3) 3/6 from $^3D_2 \to {}^3P_2$; π and σ components together; field 40,000 gauss. Again the Zn lines are more affected than the Cd. Comparison of the figures shows that in weak fields the patterns are symmetrical, while in stronger fields the positions are affected, so that the pattern is no longer symmetrical.

15. Mg, partial Paschen-Back effect in a diffuse triplet.

$^3D \to {}^3P_0$; 3829·4 Å; pattern (0)(2) 2 4/2.

$^3D \to {}^3P_1$; 3832·3 Å; pattern (0)(1)(2) 3 4 5/2.

$^3D \to {}^3P_2$; 3838·3 Å; pattern (0)(1)(2) 2 3 4/2.

π components are above and σ below; the weakest components are partly lost in reproduction.

16. Partial Paschen-Back effect in a diffuse doublet.

$^2D \to {}^2P_{1/2}$ Na 3302·3 Å; pattern (1)(2) 2 4 6/3.

$^2D \to {}^2P_{3/2}$ Na 3302·9 Å; pattern (1)(3) 2 3 4 [illegible].

π components above and σ components below.

Figs. 15 and 16 were lent by Prof. J. B. Green, and all the others by Prof. E. Back; the former have appeared in *P.R.* 1934, [illegible] and the latter in Zeeman-Effekt und Multiplettstruktur by Back and Landé.

PLATE III. MAGNETIC SPLITTING PATTERNS

1. H_α; 6563·04 A.; both π and σ components; field 39,000 g.

2. $3\,^1D \rightarrow 2\,^1P$. Cd 6438·71 A.; normal triplet (0) 1/1; π components above, σ below.

3. $^2P_{1\frac{1}{2}} \rightarrow {}^2S_{\frac{1}{2}}$; Na 5890·19 A., D_2 line on left; $^2P_{\frac{1}{2}} \rightarrow {}^2S_{\frac{1}{2}}$; Na 5895·16 A., D_1 line on right; the line in the absence of magnetic field below, π and σ components together above.

4. $2\,^3S_1 \rightarrow 2\,^3P_0$; Zn 4680·38 A.; pattern (0) 4/2.

5. $2\,^3S_1 \rightarrow 2\,^3P_1$; Zn 4722·34 A.; pattern (1) 3 4/2.

6. $2\,^3S_1 \rightarrow 2\,^3P_2$; Zn 4810·71 A.; pattern (0) 1 (2) 3 4/2.

7. (^{6}S) 4p $^7P^\circ_2 \rightarrow$ (^{6}S) 4s $^7S_3$; Cr 4289·92 A.; pattern *(0)* (1) (2) *4* 5 6 7 8/3; σ components above, π below.

8. $^7P^\circ_3 \rightarrow {}^7S_3$; Cr 4374·75 A.; pattern (1) (2) (3) 21 22 *23 24* 25 26/12. The components separated by an interval of 1/12. $o_m H$ are not perfectly resolved, but the separations of the strongest σ components can be accurately measured.

9. $^7P^\circ_4 \rightarrow {}^7S_3$; Cr 4254·49 A.; pattern *(0)* (1) (2) (3) *4* 5 6 7 8 9 10/4; the weakest components $\pm$ 10/4 do not appear in the reproduction.

10. $3\,^2D_{1\frac{1}{2}} \rightarrow 2\,^2P_{1\frac{1}{2}}$; Tl 3529·58 A.; pattern (4) 8 *(12) 16* 24/15; π components above, σ below.

11. $3\,^3D_1 \rightarrow 2\,^3P_0$; Zn 3282·42 A.; pattern (0) 1/2; lines in absence of field above, π and σ components together below. In the lower figure the forbidden transition $3\,^3D_2 \rightarrow 2\,^3P_0$, with pattern (0) 7/6 is visible, while in the upper figure it is absent.

12. 6p.6d 3° → 6p^2 ^{1}D; Pb 4062·25 A. The observed pattern *(0)* (0·367) 0·855 1·222 *1·589* is best satisfied by g factors of 0·857 and 1·222; the g values of lead are abnormal for the vector coupling is not (**LS**).

13. $^3D_2 \rightarrow {}^3P_2$; Zn 3345 A. on left; $^3D_2 \rightarrow {}^3P_2$; Cd 3610 A. on right; both patterns (2) *(4)* 5 *7 9* 11/6; π and σ components together. Both photographs were taken in a field of 40,000 g. making $o_m = 1·85$ cm.$^{-1}$; the incipient Paschen-Back effect is more obvious in Zn than in Cd, because the ^{3}D interval factor is 1·9 cm.$^{-1}$ in Zn and 6·0 in Cd.

14. $^3D_{2,1} \rightarrow {}^3P_1$; Zn 3302–3 A. on left; $^3D_{2,1} \rightarrow {}^3P_1$; Cd 3466–8 A. on right. In each element the intense pattern is *(0)* (2) *5* 7 9/6 from $^3D_2 \rightarrow {}^3P_1$ and the faint *1* (2) 3/2 from $^3D_1 \rightarrow {}^3P_1$; π and σ components together; field 40,000 g. Again the Zn lines are more distorted than the Cd. Comparison of the figures shows that in weak fields the intensities are distorted, while in stronger fields the positions are affected, so that the pattern is no longer symmetrical.

15. Mg; partial Paschen-Back effect in a diffuse triplet.

$^3D \rightarrow {}^3P_0$; 3829·51 A.; pattern *(0) 0* (2) 2 4/2.
$^3D \rightarrow {}^3P_1$; 3832·46 A.; pattern *0 (1)* 1 (2) 3 4/2.
$^3D \rightarrow {}^3P_2$; 3838·46 A.; pattern *(0)* 0 (1) 1 (2) *2* 3 4/2.

σ components are above and π below; the weakest components are partly lost in reproduction.

16. Partial Paschen-Back effect in a diffuse doublet.

$^2D \rightarrow {}^2P_{1\frac{1}{2}}$; Na 5688·20 A.; pattern *(1)* 1 (2) *2* 4 5/3.
$^2D \rightarrow {}^2P_{\frac{1}{2}}$; Na 5682·90 A.; pattern *(0)* (1) 1 (2) 2 *3* 4/3.

π components are above and σ components below.

Figs. 13 and 14 were lent by Prof. J. B. Green, and all the others by Prof. E. Back; the former have appeared in *PR*, 1934, 45, and the latter in *Zeemaneffekt und Multiplettstruktur* by Back and Landé.

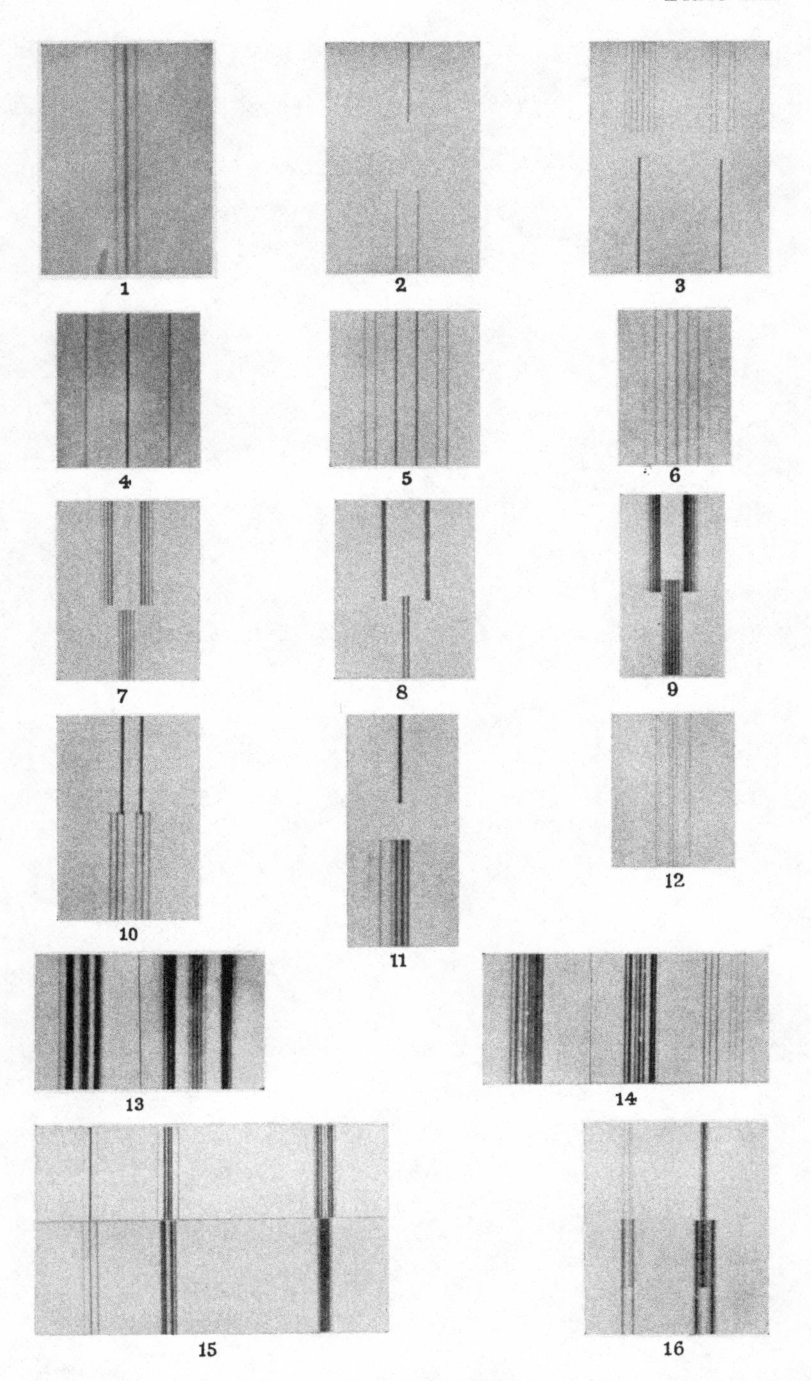
1
2
3
4
5
6
7
8
9
10
11
12
13
14
15
16

electromagnets can produce a steady field greater than this, though Kapitza using a sudden discharge obtains a much greater field for a very short time. Conversion to wave-lengths gives $\Delta\lambda = \lambda^2 o_m$, so that the interval to be measured is 1·2 A. at 8000 A., but only 0·076 A. at 2000 A.

2. Anomalous Zeeman effect.

Lines not belonging to a singlet series split in a magnetic field into more than three components. These are, however, still symmetrical about the position of the undisplaced line, and polarised parallel or perpendicular to the magnetic field.

The group of components into which a line splits are determined absolutely in number, displacement and intensity, and are said to constitute a 'Zeeman type'. Preston* found that two lines exhibit the same Zeeman type if they belong to the same series, or if they are members of different series, such as the principal and the sharp, which arise from the same term sequences or thirdly if they belong to identical series occurring in different elements. Thus the $4\,^2P_{1\frac{1}{2}} \rightarrow 1\,^2S_{\frac{1}{2}}$ line of sodium and the $5\,^2S_{\frac{1}{2}} \rightarrow 2\,^2P_{1\frac{1}{2}}$ line of caesium split in the same way when produced in a magnetic field. In terms of the vector model this clearly means that the Zeeman type is a function of the **L**, **S** and **J** vectors of the two combining terms, but not of the chief quantum number n nor of the inner electron groups.

Because the displacements of the Zeeman components depend on the strength of the magnetic field, they are conveniently expressed in some unit which varies with the field, such as the displacement of the σ components of the normal triplet. In 1907 Runge† used it, and found that the displacements of all Zeeman components can be written as simple fractions having small denominators. Thus the displacements of the π and σ components of the D_1 line of sodium are respectively $\pm\frac{2}{3}$ and $\pm\frac{4}{3}$ Lorentz units. This is commonly abbreviated to read $\frac{(2)\,4}{3}$, where all π components are enclosed in brackets. Or again the displacements

* Preston, *Phil. Trans. R.S. Dublin*, 1899, **7** 7.

† Runge, *PZ*, 1907, **8** 232.

may be written in decimals as in Fig. 6·2, which gives the displacements of the components of the $^2P_{1\frac{1}{2}} \rightarrow {}^2S_{\frac{1}{2}}$ line in various spectra, the field being 30,000 gauss. In this field the Lorentz

Element	Wave-length	Displacements	
		In A.	In cm.$^{-1}$
Caesium	8521	±(0·341), 1·024, 1·705	±(0·470), 1·410, 2·348
Sodium	5890	±(0·166), 0·498, 0·830	±(0·479), 1·435, 2·388
Copper	3247	±(0·049), 0·149, 0·247	±(0·465), 1·413, 2·342
Zinc+	2026	±(0·019), 0·058, 0·097	±(0·46), 1·41, 2·36

Fig. 6·2. Splitting of the $^2P_{1\frac{1}{2}} \rightarrow {}^2S_{\frac{1}{2}}$ line in four spectra; the field is 30,000 gauss and $o_m = 1{\cdot}409$ cm.$^{-1}$

interval is 1·409 cm.$^{-1}$, so that the displacements of the Cu 3247 A. line are $0{\cdot}33o_m$, $1{\cdot}00o_m$ and $1{\cdot}66o_m$; clearly then the Runge fraction is $\frac{(1)\,3\;5}{3}$. Examples of other Zeeman types with their appropriate Runge fractions are shown in Fig. 6·1, the brightest π and σ components being in italics.

The importance of these laws in the analysis of a spectrum can hardly be exaggerated, for the Zeeman type indicates at once the term combination from which a line arises.

3. The Quantum theory

Nowhere perhaps have theory and experiment worked more closely together than in the construction of modern spectral theory, and nowhere is this more easily illustrated than in a discussion of the Zeeman effect.

The wave-number ν of a spectral line, produced when an atom jumps from a stationary state having energy E' to a second having energy E'', is given by the relation

$$ch\nu = E' - E''.$$

When a magnetic field is applied, the energies of these states may be supposed to change by $\Delta E'$ and $\Delta E''$, so that the wave-number of the line emitted changes by $\Delta\nu$, where

$$ch\,\Delta\nu = \Delta E' - \Delta E''. \qquad \text{......(6·2)}$$

If these increments of energy could assume a continuous series of values, the line would simply broaden, but as in fact the line splits into a number of components the increments of energy must themselves be restricted by quantum conditions.*

The above argument assumes only the existence of stationary states in the atom; to proceed further consider an electron revolving in a circle about a positively charged core; the quantum restriction of angular momentum may then be shown to restrict also the magnetic moment of the atom, for the classical theory shows that one is a constant multiple of the other. Thus the magnetic moment is

$$\frac{iA}{c} = \frac{\omega e}{2\pi} \cdot \frac{\pi r^2}{c} = \frac{e}{2m_e c} \cdot m_e \omega r^2, \qquad \text{......(6·3)}$$

where the current i and the electronic charge e are both measured in e.s.u., A is the area of the orbit and ω the angular velocity. Though the above equations apply only to a circular orbit the result is equally valid for an ellipse, so that if the angular momentum of the atom can only assume the values $J \cdot \frac{h}{2\pi}$, the magnetic moment is restricted to the values

$$J \cdot \frac{e}{2m_e c} \cdot \frac{h}{2\pi} = J \cdot \frac{eh}{4\pi m_e c}, \qquad \text{......(6·4)}$$

Thus $\frac{eh}{4\pi m_e c}$ is the quantum unit of magnetic moment; it is called the Bohr magneton, is commonly written μ_B, and is numerically $9{\cdot}156 \,.\, 10^{-21}$ erg gauss^{-1}.

That the magnetic moment is quantised does not indicate the energy of each state in the magnetic field, but as experiment leaves no doubt that the magnetic increments of energy are proportional to the field H, the product $J\mu_B H$ is obviously suggested, for this at least has the right dimensions; and in fact the wave mechanics justifies this guess when **J** is parallel to **H**; more generally the energy increment is $J \cos(\mathbf{JH})\, \mu_B H$, since the orbital area projected on a plane perpendicular to H is $A \cos(\mathbf{JH})$, when **J** is inclined to **H**.

Since the Zeeman lines are sharp the projection of **J** on the

* Van Lohuizen, *K. Akad. Amsterdam Proc.* 1919, 22 190.

magnetic axis must be quantised, and to specify this we introduce a magnetic quantum number M defined by the relation

$$M = J\cos(\mathbf{JH}), \qquad \text{......(6·5)}$$

so that M is restricted by the condition

$$J \geqslant M \geqslant -J$$

and can assume $(2J+1)$ between these two extremes. The orientations permitted when $J=3$ are shown graphically in Fig. 6·3.

Thus in general the energy increment is

$$\Delta E = J\cos(\mathbf{JH})\,\mu_B H$$
$$= M\mu_B H \text{ ergs}; \qquad \text{......(6·6)}$$

but the energy is often more conveniently expressed in wave-numbers, so divide by ch and write

$$\Delta E/ch = Mo_m, \qquad \text{......(6·7)}$$

where o_m is defined by

$$o_m = \frac{\mu_B H}{ch} = \frac{eH}{4\pi m_e c^2} = 4{\cdot}665\,.\,10^{-5}\,H\ \text{cm.}^{-1} \qquad \text{...(6·8)}$$

The above theory assumes that a magnetic field does not destroy the coupling which links the spin and orbital vectors, and

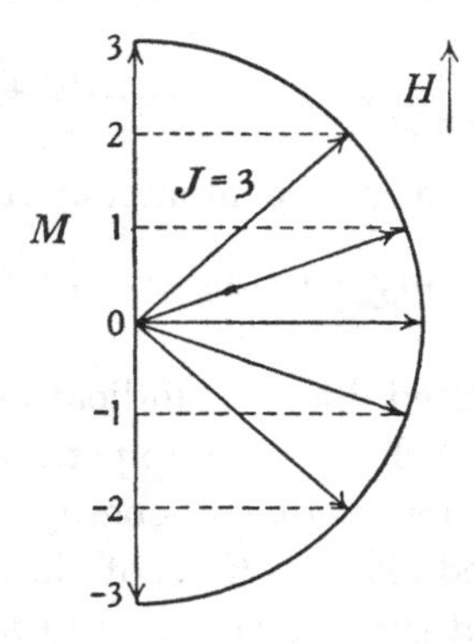

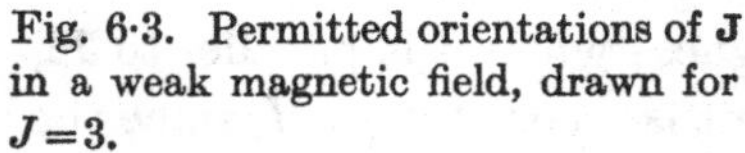
Fig. 6·3. Permitted orientations of **J** in a weak magnetic field, drawn for $J=3$.

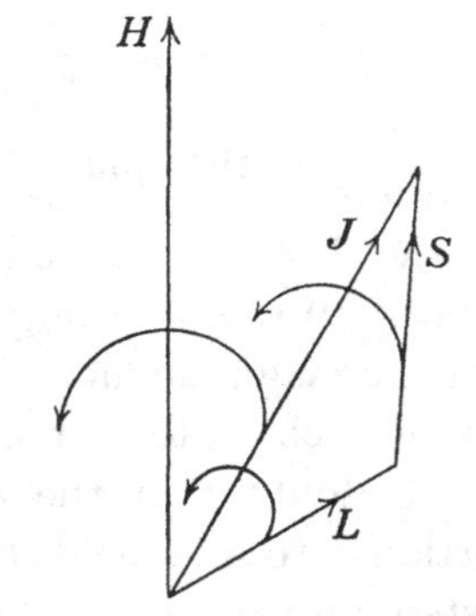

Fig. 6·4. In a weak magnetic field **J** precesses round the direction of the field **H**, while **L** and **S** precess round **J**.

that while **J** precesses round **H**, **L** and **S** precess together round **J** (Fig. 6·4). Experiment shows that this assumption is valid when the field is weak, but that when it grows stronger it breaks the coupling of **L** and **S**, and these then precess independently round **H**; this changed motion produces a changed splitting of the spectral lines, a splitting commonly known as the Paschen-Back effect.

Applied to a line of a singlet series, this theory must explain the normal triplet which experiment reveals. In a magnetic field the wave-number changes by

$$\Delta\nu = \Delta E'/ch - \Delta E''/ch,$$

while the change in the energy of a single state is

$$\Delta E/ch = Mo_m. \qquad \ldots\ldots(6\cdot7)$$

So that combining these

$$\Delta\nu = (M' - M'')\,o_m. \qquad \ldots\ldots(6\cdot9)$$

This equation suggests that the $^1D_2 \to {}^1P_1$ line will split into seven components with displacements 0, $\pm o_m$, $\pm 2o_m$, $\pm 3o_m$; whereas in fact experiment gives only three. To obtain agreement Rubinowicz and Bohr* put forward the selection rule

$$\Delta M = 0 \quad \text{or} \quad \pm 1;$$

while to account for the empirical polarisation they postulated that a transition gives rise to a π component when $\Delta M = 0$, and to a σ component when $\Delta M = \pm 1$. This is very simply shown in Fig. 6·5, where arrows show the jumps allowed. Vertical arrows indicate π components, and oblique arrows σ components.

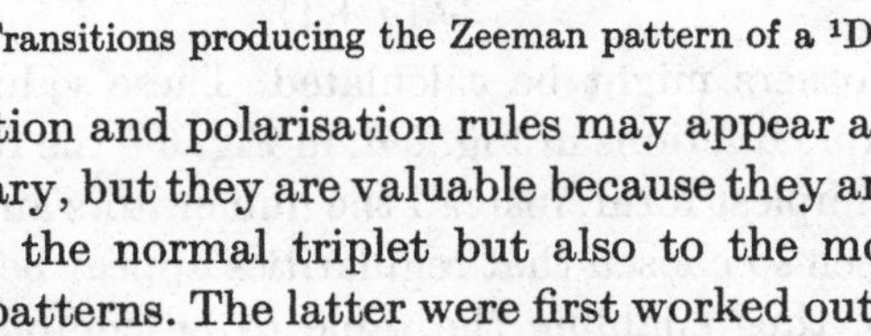

Term	M
$^1D_2$	−2 −1 0 1 2
$^1P_1$	−1 0 1
$(M' - M'')$	(0), ±1

Fig. 6·5. Transitions producing the Zeeman pattern of a $^1D_2 \to {}^1P_1$ line.

The selection and polarisation rules may appear at first somewhat arbitrary, but they are valuable because they are applicable not only to the normal triplet but also to the more complex anomalous patterns. The latter were first worked out empirically by Landé, and though a satisfactory theory has since been con-

* Rubinowicz, *PZ*, 1918, **19** 441, 465; Bohr, *Kopenhagener Akad.* 1918, **14** 1.

structured, the empirical approach will perhaps bring out the physical principles more clearly. The simple theory shows that a magnetic field changes the energy of a stationary state by

$$\Delta E/ch = Mo_m. \qquad \text{......(6·7)}$$

As this did not suffice to account for the anomalous effect Landé introduced an arbitrary factor g, and wrote

$$\Delta E/ch = Mgo_m. \qquad \text{......(6·10)}$$

The splitting factor g might here be a function of M; fortunately, however, experiment shows that it depends only on L, S and J.

Term	L \ J	0	1	2	3	4	$\frac{1}{2}$	$1\frac{1}{2}$	$2\frac{1}{2}$	$3\frac{1}{2}$	$4\frac{1}{2}$	J / Term
		Singlets ($S=0$)					Doublets ($S=\frac{1}{2}$)					
S	0	$\frac{0}{0}$					2					S
P	1		1				$\frac{2}{3}$	$\frac{4}{3}$				P
D	2			1				$\frac{4}{5}$	$\frac{6}{5}$			D
F	3				1				$\frac{6}{7}$	$\frac{8}{7}$		F
		Triplets ($S=1$)					Quartets ($S=1\frac{1}{2}$)					
S	0		2					2				S
P	1	$\frac{0}{0}$	$\frac{3}{2}$	$\frac{3}{2}$			$\frac{8}{3}$	$\frac{26}{15}$	$\frac{8}{5}$			P
D	2		$\frac{1}{2}$	$\frac{7}{6}$	$\frac{4}{3}$		0	$\frac{6}{5}$	$\frac{48}{35}$	$\frac{10}{7}$		D
F	3			$\frac{2}{3}$	$\frac{13}{12}$	$\frac{5}{4}$		$\frac{2}{5}$	$\frac{36}{35}$	$\frac{26}{21}$	$\frac{4}{3}$	F

Fig. 6·6. The magnetic splitting factor, g.

With this assumption Landé,* was able to work out many values of g empirically, and then to devise a formula,

$$g = 1 + \frac{J(J+1) + S(S+1) - L(L+1)}{2J(J+1)}, \qquad \text{......(6·11)}$$

from which others might be calculated. These values of g are shown as simple fractions in Fig. 6·6; in Fig. 6·8 the fractions are not in their simplest form; instead the numerators and denominators have been so chosen that regularities appear between rows and columns, thus enabling the table to be compiled without applying the formula to each term.†

* Landé, *ZP*, 1923, **15** 189. † Kiess and Meggers, *BSJ*, 1928, **1** 641.

As an example of the application of this formula, consider the brightest line of a sharp triplet $^3S_1 \rightarrow {}^3P_2$. For $^3S_1$

$$g = 1 + \frac{2+2-0}{2.2} = 2,$$

while for $^3P_2$ $\qquad g = 1\tfrac{1}{2}.$

And from these values, with the aid of the table given in Fig. 6·7 the splitting of the line can be deduced as

$$\frac{(0)\,(1)\,2\,3\,4}{2}.$$

Term	g	M	-2	-1	0	1	2
$^3S_1$	2	$M_i g_i$		-2	0	2	
$^3P_2$	$\frac{3}{2}$	$M_f g_f$	-3	$-\frac{3}{2}$	0	$\frac{3}{2}$	3
$M_i g_i - M_f g_f$			$\pm\frac{(0)\,(1)\,2\,3\,4}{2}$				

Term	g	M	-2	-1	0	1	2
$^3P_2$	$\frac{3}{2}$	$M_i g_i$	-3	$-1\frac{1}{2}$	0	$1\frac{1}{2}$	3
$^3D_2$	$\frac{7}{6}$	$M_f g_f$	$-2\frac{1}{3}$	$-1\frac{1}{6}$	0	$1\frac{1}{6}$	$2\frac{1}{3}$
$M_i g_i - M_f g_f$			$\pm\frac{(2)\,(4)\,5\,7\,9\,11}{6}$				

Term	g	M	$-1\frac{1}{2}$	$-\frac{1}{2}$	$\frac{1}{2}$	$1\frac{1}{2}$
$^4P_{1\frac{1}{2}}$	$\frac{26}{15}$	$M_i g_i$	$-\frac{13}{5}$	$-\frac{13}{15}$	$\frac{13}{15}$	$\frac{13}{5}$
$^4D_{1\frac{1}{2}}$	$\frac{6}{5}$	$M_f g_f$	$-\frac{9}{5}$	$-\frac{3}{5}$	$\frac{3}{5}$	$\frac{9}{5}$
$M_i g_i - M_f g_f$			$\pm\frac{(4)\,(12)\,14\,22\,30}{15}$			

Fig. 6·7. Calculation of the Zeeman patterns of the three lines, $^3S_1 \rightarrow {}^3P_2$, $^3P_2 \rightarrow {}^3D_2$, $^4P_{1\frac{1}{2}} \rightarrow {}^4D_{1\frac{1}{2}}$.

The figure also shows similar calculations for triplet and quartet lines.

Alternatively the Zeeman type can be deduced from a level diagram, such as Fig. 6·9, which shows a $^2D_{1\frac{1}{2}} \rightarrow {}^2P_{1\frac{1}{2}}$ combination.

Term	L \ J	0	1	2	3	4	5	6	7	Term
							Singlets ($S=0$)			
S	0	$\frac{0}{0}$								S
P	1		$\frac{2}{2}$							P
D	2			$\frac{6}{6}$						D
F	3				$\frac{12}{12}$					F
G	4					$\frac{20}{20}$				G
							Triplets ($S=1$)			
S	0		$\frac{4}{2}$							S
P	1	$\frac{0}{0}$	$\frac{3}{2}$	$\frac{9}{6}$						P
D	2		$\frac{1}{2}$	$\frac{7}{6}$	$\frac{16}{12}$					D
F	3			$\frac{4}{6}$	$\frac{13}{12}$	$\frac{25}{20}$				F
G	4				$\frac{9}{12}$	$\frac{21}{20}$	$\frac{36}{30}$			G
							Quintets ($S=2$)			
S	0			$\frac{12}{6}$						S
P	1		$\frac{5}{2}$	$\frac{11}{6}$	$\frac{20}{12}$					P
D	2	$\frac{0}{0}$	$\frac{3}{2}$	$\frac{9}{6}$	$\frac{18}{12}$	$\frac{30}{20}$				D
F	3		$\frac{0}{2}$	$\frac{6}{6}$	$\frac{15}{12}$	$\frac{27}{20}$	$\frac{42}{30}$			F
G	4			$\frac{2}{6}$	$\frac{11}{12}$	$\frac{23}{20}$	$\frac{38}{30}$	$\frac{56}{42}$		G
							Septets ($S=3$)			
S	0				$\frac{24}{12}$					S
P	1			$\frac{14}{6}$	$\frac{23}{12}$	$\frac{35}{20}$				P
D	2		$\frac{6}{2}$	$\frac{12}{6}$	$\frac{21}{12}$	$\frac{33}{20}$	$\frac{48}{30}$			D
F	3	$\frac{0}{0}$	$\frac{3}{2}$	$\frac{9}{6}$	$\frac{18}{12}$	$\frac{30}{20}$	$\frac{45}{30}$	$\frac{63}{42}$		F
G	4		$-\frac{1}{2}$	$\frac{5}{6}$	$\frac{14}{12}$	$\frac{26}{20}$	$\frac{41}{30}$	$\frac{59}{42}$	$\frac{80}{56}$	G

Fig. 6·8. The magnetic splitting factor, g. The fractions
the trouble of calculating

Term	L \ J	1/2	1 1/2	2 1/2	3 1/2	4 1/2	5 1/2	6 1/2	7 1/2	8 1/2	Term
									Doublets ($S=\frac{1}{2}$)		
S	0	6/3									S
P	1	2/3	20/15								P
D	2		12/15	42/35							D
F	3			30/35	72/63						F
G	4				56/63	110/99					G
									Quartets ($S=1\frac{1}{2}$)		
S	0		30/15								S
P	1	8/3	26/15	56/35							P
D	2	0/3	18/15	48/35	90/63						D
F	3		6/15	36/35	78/63	132/99					F
G	4			20/35	62/63	116/99	182/143				G
									Sextets ($S=2\frac{1}{2}$)		
S	0			70/35							S
P	1		36/15	66/35	108/63						P
D	2	10/3	28/15	58/35	100/63	154/99					D
F	3	−2/3	16/15	46/35	88/63	142/99	208/143				F
G	4		0/15	30/35	72/63	126/99	192/143	270/195			G
									Octets ($S=3\frac{1}{2}$)		
S	0				126/63						S
P	1			80/35	122/63	176/99					P
D	2		42/15	72/35	114/63	168/99	234/143				D
F	3	12/3	30/15	60/35	102/63	156/99	222/143	300/195			F
G	4	−4/3	14/15	44/35	86/63	140/99	206/143	284/195	374/255		G

are so chosen that the table can be constructed without each factor separately.

The continuous lines indicate σ components, the dotted lines π components.

A table of some patterns is added (p. 95, Fig. 6·10).

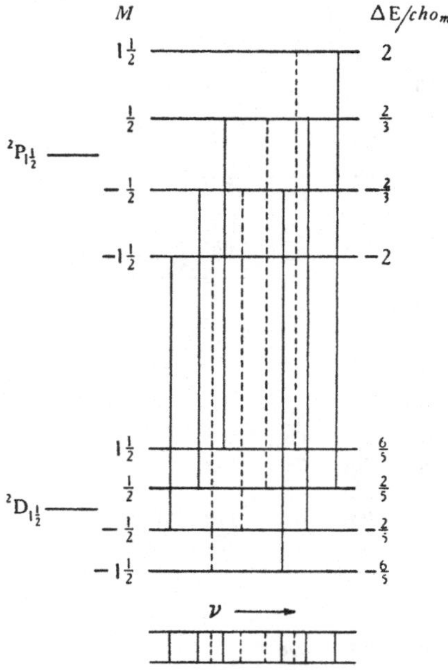

Fig. 6·9. Transitions producing the Zeeman pattern of a $^2P_{1\frac{1}{2}} \rightarrow {}^2D_{1\frac{1}{2}}$ line. The π components are dotted, the σ components drawn full.

4. The Spinning electron

The old quantum theory thus gives an admirable account of the Zeeman effect provided that the empirical splitting factor g is accepted without cavil; but to press this point is to reveal the elementary theory as inadequate.

The Landé splitting factor measures the ratio of the magnetic to the mechanical moment of an atom, on a scale which will make the g factor of a point charge moving in an ellipse unity. Now the g factor is unity only in the terms of a singlet series; in these terms the spin vector is zero, so that the 'anomalous' g values may be reasonably attributed to the electron; and as the g value of every term having $L=0$ is 2, the electron may be assumed to have twice as great a magnetic moment as the

Doublet systems.

$^2P_{\frac{1}{2}}-^2D_{1\frac{1}{2}}$	(0·07), 0·73, *0·87*
$^2P_{1\frac{1}{2}}-^2D_{1\frac{1}{2}}$	(0·27, *0·80*), 0·53, *1·07*, 1·60
$^2P_{1\frac{1}{2}}-^2D_{2\frac{1}{2}}$	(*0·07*, 0·20), *1·00*, 1·13, 1·27, 1·40
$^2D_{1\frac{1}{2}}-^2F_{2\frac{1}{2}}$	(*0·03*, 0·09), 0·77, 0·83, 0·89, *0·94*
$^2D_{2\frac{1}{2}}-^2F_{2\frac{1}{2}}$	(0·17, 0·51, *0·86*), 0·34, 0·69, *1·03*, 1·37, 1·71
$^2D_{2\frac{1}{2}}-^2F_{3\frac{1}{2}}$	(*0·03*, 0·09, 0·14), *1·00*, 1·06, 1·11, 1·17, 1·23, 1·29

Triplet systems.

$^3P_0-^3D_1$	(0·00), 0·50
$^3P_1-^3D_1$	(1·00), 0·50, 1·50
$^3P_1-^3D_2$	(*0·00*, 0·33), *0·83*, 1·17, 1·50
$^3P_2-^3D_1$	(*0·00*, 1·00), 0·50, 1·50, *2·50*
$^3P_2-^3D_2$	(0·33, *0·67*), 0·83, *1·17*, *1·50*, 1·83
$^3P_2-^3D_3$	(*0·00*, 0·17, 0·33), *1·00*, 1·17, 1·33, 1·50, 1·67
$^3D_1-^3F_2$	(*0·00*, 0·17), 0·50, 0·67, *0·83*
$^3D_2-^3F_2$	(0·50, *1·00*), 0·17, *0·67*, *1·17*, 1·67
$^3D_2-^3F_3$	(*0·00*, 0·08, 0·17), *0·92*, 1·00, 1·08, 1·17, 1·25
$^3D_3-^3F_2$	(*0·00*, 0·67, 1·33), 0·00, 0·67, 1·33, 2·00, *2·67*
$^3D_3-^3F_3$	(0·25, 0·50, *0·75*), 0·58, 0·83, *1·08*, *1·33*, 1·58, 1·83
$^3D_3-^3F_4$	(*0·00*, 0·08, 0·17, 0·25), *1·00*, 1·08, 1·17, 1·25, 1·33, 1·42, 1·50

Quartet systems.

$^4S_{1\frac{1}{2}}-^4P_{\frac{1}{2}}$	(0·33), *1·67*, 2·33
$^4S_{1\frac{1}{2}}-^4P_{1\frac{1}{2}}$	(0·13, *0·40*), 1·60, *1·87*, 2·13
$^4S_{1\frac{1}{2}}-^4P_{2\frac{1}{2}}$	(*0·20*, 0·60), *1·00*, 1·40, 1·80, 2·20
$^4P_{\frac{1}{2}}-^4D_{\frac{1}{2}}$	(1·33), 1·33
$^4P_{\frac{1}{2}}-^4D_{1\frac{1}{2}}$	(0·73), *0·47*, 1·93
$^4P_{1\frac{1}{2}}-^4D_{\frac{1}{2}}$	(0·87), 0·87, *2·60*
$^4P_{1\frac{1}{2}}-^4D_{1\frac{1}{2}}$	(0·27, *0·80*), 0·93, *1·47*, 2·00
$^4P_{1\frac{1}{2}}-^4D_{2\frac{1}{2}}$	(*0·18*, 0·54), *0·83*, 1·19, 1·55, 1·91
$^4P_{2\frac{1}{2}}-^4D_{1\frac{1}{2}}$	(*0·20*, 0·60), 1·00, 1·40, 1·80, *2·20*
$^4P_{2\frac{1}{2}}-^4D_{2\frac{1}{2}}$	(0·11, 0·34, *0·57*), 1·03, 1·26, *1·49*, 1·71, 1·94
$^4P_{2\frac{1}{2}}-^4D_{3\frac{1}{2}}$	(*0·09*, 0·26, 0·43), *1·00*, 1·17, 1·34, 1·51, 1·69, 1·86

Fig. 6·10. Zeeman patterns of doublet, triplet and quartet lines.

The π components are enclosed in parenthesis. The brightest π and σ components are italicised.

A complete table of theoretical patterns, going up to ^8I terms and containing inter-system lines, has been given by Kiess and Meggers, *BSJ*, 1928, **1** 641.

Lorentz theory would predict for a rotating point charge. Thus if the angular momentum of an electron is $\frac{1}{2} \cdot \frac{h}{2\pi}$, the magnetic moment must be

$$2 \cdot \frac{e}{2m_e c} \cdot \frac{1}{2} \cdot \frac{h}{2\pi} \qquad \text{......(6·12)}$$

or $\frac{eh}{4\pi m_e c}$, and not $\frac{eh}{8\pi m_e c}$ as the Lorentz theory would indicate. This magnetic moment is inherent in the structure of the electron itself, and does not prevent it from developing a further magnetic moment as it revolves about the nucleus. But whereas the inherent magnetic moment of a group of electrons is proportional to S, the orbital magnetic moment is proportional to L.

When **L** and **S** are parallel, the combination of the two magnetic moments presents no difficulties. Thus Fig. 6·11 is drawn for any term such as $^2P_{1\frac{1}{2}}$, in which $J = L + S$; the projections of **L**, **S** and **J** on the axis of the magnetic field are written $\mathbf{M_L}$, $\mathbf{M_S}$ and **M** respectively, so that

$$\frac{M_L}{L} = \frac{M_S}{S} = \frac{M}{J}.$$

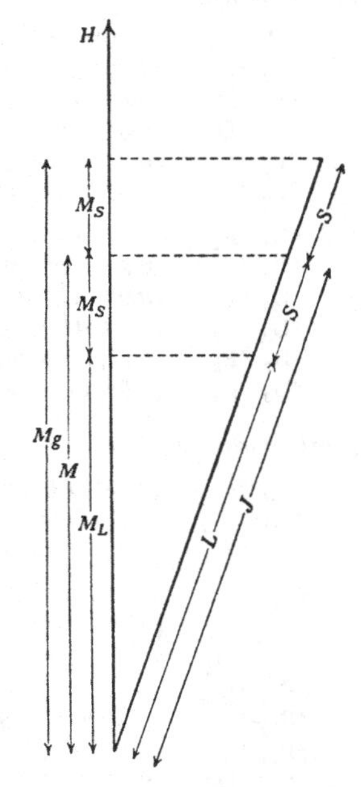

Fig. 6·11. Derivation of the magnetic splitting factor of an atomic state in which the orbital and spin vectors are parallel.

Because of the inherent magnetism of the electron the component of the magnetic moment will be $\mathbf{M_L} + 2\mathbf{M_S}$ instead of the $\mathbf{M_L} + \mathbf{M_S}$ or **M** which Lorentz predicted. But

$$M_L + 2M_S = M + M_S = M\left(1 + \frac{S}{J}\right),$$

so that
$$g = 1 + \frac{S}{J}.$$

In the $^2P_{1\frac{1}{2}}$ term this gives $g = \frac{4}{3}$ as experiment requires; and a glance at Landé's table of g values, Fig. 6·6, shows that whenever

$\mathbf{J}=\mathbf{L}+\mathbf{S}$, g is always $(1+S/J)$, while when $\mathbf{J}=\mathbf{L}-\mathbf{S}$, g is $(1-S/J)$.

When the orbital and spin vectors are not parallel, the magnetic splitting factor cannot be obtained by simple geometry; if, however, the energy of interaction of two vectors is assumed proportional to the cosine of the angle between them, a formula results which has only to be 'refined' by the quantum mechanics to agree with experiment. Equations (6·5) and (6·7) contain the results of the simple theory, which ignores the intrinsic magnetism of the electron; combined these equations show that

$$\Delta E/ch = o_m J \cos(\mathbf{JH}). \qquad \text{......(6·13)}$$

Accordingly as $\mathbf{J}$ is the vectorial sum of $\mathbf{L}$ and $\mathbf{S}$, and the g factor of the electron is 2, the displacements of a Zeeman term should be

$$\Delta E/ch = o_m \{L \overline{\cos(\mathbf{LH}) + 2S \cos(\mathbf{SH})}\}. \qquad \text{......(6·14)}$$

In a weak field the orbital vector precesses round the electronic vector $\mathbf{J}$, while $\mathbf{J}$ precesses round $\mathbf{H}$ (Fig. 6·4); accordingly the mean value of $\cos(\mathbf{LH})$ has to be taken, as is indicated in the equation by a superposed bar. The mean value is

$$\overline{\cos(\mathbf{LH})} = \cos(\mathbf{LJ})\cos(\mathbf{JH}).$$

And what is true of L is true also of S, so that

$$\Delta E/ch = o_m J \cos(\mathbf{JH}) \left\{ \frac{L}{J}\cos(\mathbf{LJ}) + \frac{2S}{J}\cos(\mathbf{SJ}) \right\}. \qquad \text{......(6·15)}$$

In macroscopic physics

$$\frac{L}{J}\cos(\mathbf{LJ}) = \frac{J^2+L^2-S^2}{2J^2},$$

but the quantum mechanics shows that in dealing with atoms this must be refined to

$$\frac{L}{J}\cos(\mathbf{LJ}) = \frac{J(J+1)+L(L+1)-S(S+1)}{2J(J+1)}, \qquad \text{......(6·16)}$$

while $$\frac{S}{J}\cos(\mathbf{SJ}) = \frac{J(J+1)+S(S+1)-L(L+1)}{2J(J+1)}.$$

Thus $$\Delta E/ch = o_m M \left\{1 + \frac{J(J+1) + S(S+1) - L(L+1)}{2J(J+1)}\right\}, \quad \text{......(6·17)}$$

where $$M = J \cos(\mathbf{JH}).$$

This is the formula found empirically by Landé, and expressed in equations 6·10 and 6·11 above.

5. Intensity rules

The brightest π and σ components of any Zeeman type can be predicted by two simple rules.

In the combination of two terms having the same value of J, or, which is identical, having the same number of Zeeman components, those π components are brightest which arise in arrows at the extremities of the usual scheme, while the brightest σ components arise in arrows near the centre. In particular the π component arising from the transition from $M=0$ to $M=0$ is always missing.

But when the two combining terms have different values of J, the bright π components arise in arrows near the centre, while the bright σ components lie at the extremities (Fig. 6·7).

The more precise intensity relations which have been obtained by use of the correspondence principle are interesting and are discussed in a later chapter, but in actual fact these two simple rules have sufficed for the analysis of all Zeeman types, for they make clear four intensity classes, and the class to which an unknown Zeeman type belongs determines certain facts about the J and g values of the terms producing it.

Suppose that an observed pattern arises from two unknown terms x and y, and that these are distinguished by the condition $J_x \leqslant J_y$; then the only possible intensity relations are the four given below, and each has its particular significance. The strongest line in each group is underlined.

Class I. The intensities are represented by

(a) $$\underline{\sigma}\,\sigma\,\sigma \ldots \qquad \ldots \pi\,\underline{\pi}\,\pi \ldots \qquad \ldots \sigma\,\sigma\,\underline{\sigma}$$

for odd multiplicity, or by

(b) $$\underline{\sigma}\,\sigma\,\sigma \ldots \qquad \ldots \pi\,\underline{\pi}\,\underline{\pi}\,\pi \ldots \qquad \ldots \sigma\,\sigma\,\underline{\sigma}$$

for even multiplicity.

A line belongs to this class only when $J_x < J_y$ and $g_x < g_y$, examples being $^3P_2\,^3D_1$ and $^2P_{\frac{1}{2}}\,^2D_{1\frac{1}{2}}$ of Fig. 6·1.

Class II. The intensities are represented by

(*a*) $\ldots\ \sigma\ \sigma\ \underline{\sigma} \qquad \ldots\ \pi\ \underline{\pi}\ \pi\ \ldots \qquad \underline{\sigma}\ \sigma\ \sigma\ \ldots$

for odd multiplicity, or by

(*b*) $\ldots\ \sigma\ \sigma\ \underline{\sigma} \qquad \ldots\ \pi\ \underline{\pi}\ \underline{\pi}\ \pi\ \ldots \qquad \underline{\sigma}\ \sigma\ \sigma\ \ldots$

for even multiplicity. A pattern belongs to this class only when $J_x < J_y$ and $g_x > g_y$. Examples of this are shown in $^3P_2\,^3D_3$, $^3P_1\,^3D_2$ and $^2P_{1\frac{1}{2}}\,^2D_{2\frac{1}{2}}$ of Fig. 6·1.

Class III. The intensities are represented by

(*a*) $\ldots\ \sigma\ \sigma\ \underline{\sigma}\ \underline{\sigma}\ \sigma\ \sigma\ \ldots \qquad \underline{\pi}\ \pi\ \pi\ \pi\ \underline{\pi} \qquad \ldots\ \sigma\ \sigma\ \underline{\sigma}\ \underline{\sigma}\ \sigma\ \sigma\ \ldots$

for odd multiplicity, or by

(*b*) $\ldots\ \sigma\ \sigma\ \underline{\sigma}\ \sigma\ \sigma\ \ldots \qquad \underline{\pi}\ \pi\ \pi\ \underline{\pi} \qquad \ldots\ \sigma\ \sigma\ \underline{\sigma}\ \sigma\ \sigma\ \ldots$

for even multiplicity. A pattern belongs to this class only when $J_x = J_y$ and $g_x \neq g_y$. Again, examples may be found in Fig. 6·1.

Class IV. A normal Zeeman triplet, consisting of one π and two σ components

$$\sigma \qquad \pi \qquad \sigma.$$

This pattern may arise when either $g_x = g_y$ or when $J_x = 0$. In complex spectra patterns apparently of this type occur when L_x, L_y are large, and g_x, g_y differ little from unity.

6. Term analysis

Every spectral line carries an identification disc, which can be read in a magnetic field, and which states in unequivocal language the terms x and y from which the line originates. From the Zeeman type one can deduce both the J values, J_x and J_y, and the g values g_x and g_y. This done one has only to turn to the table of g values given in Fig. 6·12 to identify the terms themselves.

But if the language is unequivocal it is not always legible, for when the separation of two Zeeman components is less than $o_m/5$, they cannot be fully resolved if the source is in air; and this is reduced only to $o_m/10$ if the source is *in vacuo*. Certain regularities, common to all Zeeman multiplets, may however be used to interpret words, whose letters one cannot read.

The first step is to measure up the unknown Zeeman multiplet in Lorentz units. The scale might be fixed by measuring the

g	Terms	g	Terms	g	Terms
−1·333	${}^8G_{\frac12}$	1·111	${}^2G_{4\frac12}$	1·434	${}^6F_{4\frac12}$
−0·667	${}^6F_{\frac12}$	1·133	${}^4H_{5\frac12}, {}^8I_{5\frac12}$	1·441	${}^8G_{5\frac12}$
−0·400	${}^8H_{1\frac12}$	1·143	${}^2F_{3\frac12}, {}^6G_{3\frac12}$	1·455	${}^6F_{5\frac12}$
0·000	${}^4D_{\frac12}, {}^6G_{1\frac12}, {}^8I_{2\frac12}$	1·159	${}^6I_{6\frac12}$	1·456	${}^8G_{6\frac12}$
0·286	${}^6H_{2\frac12}$	1·172	${}^4G_{4\frac12}$	1·467	${}^8G_{7\frac12}$
0·400	${}^4F_{1\frac12}$	1·200	${}^2D_{2\frac12}, {}^4D_{1\frac12}, {}^4I_{7\frac12}$	1·538	${}^8F_{6\frac12}$
0·444	${}^6I_{3\frac12}$	1·203	${}^6H_{5\frac12}$	1·552	${}^8F_{5\frac12}$
0·571	${}^4G_{2\frac12}$	1·212	${}^8H_{4\frac12}$	1·556	${}^6D_{4\frac12}$
0·667	${}^2P_{\frac12}, {}^4H_{3\frac12}, {}^8I_{3\frac12}$	1·231	${}^4H_{6\frac12}, {}^8I_{6\frac12}$	1·576	${}^8F_{4\frac12}$
0·686	${}^8H_{2\frac12}$	1·238	${}^4F_{3\frac12}$	1·587	${}^6D_{3\frac12}$
0·727	${}^4I_{4\frac12}$	1·239	${}^6I_{7\frac12}$	1·600	${}^4P_{2\frac12}$
0·800	${}^2D_{1\frac12}$	1·257	${}^8G_{2\frac12}$	1·619	${}^8F_{3\frac12}$
0·825	${}^6H_{3\frac12}$	1·273	${}^4G_{5\frac12}, {}^6G_{4\frac12}$	1·636	${}^8D_{5\frac12}$
0·828	${}^6I_{4\frac12}$	1·282	${}^6H_{6\frac12}$	1·657	${}^6D_{2\frac12}$
0·857	${}^2F_{2\frac12}, {}^6G_{2\frac12}$	1·294	${}^6I_{8\frac12}, {}^8I_{7\frac12}$	1·697	${}^8D_{4\frac12}$
0·889	${}^2G_{3\frac12}$	1·301	${}^8H_{5\frac12}$	1·714	${}^6P_{3\frac12}, {}^8F_{2\frac12}$
0·909	${}^2H_{4\frac12}$	1·314	${}^6F_{2\frac12}$	1·733	${}^4P_{1\frac12}$
0·923	${}^2I_{5\frac12}$	1·333	${}^2P_{1\frac12}, {}^4F_{4\frac12}, {}^6H_{7\frac12}$	1·778	${}^8P_{4\frac12}$
0·933	${}^8G_{1\frac12}$	1·337	${}^8I_{8\frac12}$	1·810	${}^8D_{3\frac12}$
0·965	${}^4I_{5\frac12}$	1·343	${}^6G_{5\frac12}$	1·867	${}^6D_{1\frac12}$
0·970	${}^4H_{4\frac12}, {}^8I_{4\frac12}$	1·354	${}^8H_{6\frac12}$	1·886	${}^6P_{2\frac12}$
0·984	${}^4G_{3\frac12}$	1·365	${}^8G_{3\frac12}$	1·937	${}^8P_{3\frac12}$
1·029	${}^4F_{2\frac12}$	1·368	${}^8I_{9\frac12}$	2·000	${}^2S_{\frac12}, {}^4S_{1\frac12}, {}^6S_{2\frac12}, {}^8S_{3\frac12}, {}^8F_{1\frac12}$
1·035	${}^6I_{5\frac12}$	1·371	${}^4D_{2\frac12}$	2·057	${}^8D_{2\frac12}$
1·048	${}^8H_{3\frac12}$	1·385	${}^6G_{6\frac12}$	2·286	${}^8P_{2\frac12}$
1·067	${}^6F_{1\frac12}$	1·388	${}^8H_{7\frac12}$	2·400	${}^6P_{1\frac12}$
1·071	${}^6H_{4\frac12}$	1·397	${}^6F_{3\frac12}$	2·667	${}^4P_{\frac12}$
1·077	${}^2I_{6\frac12}$	1·412	${}^8H_{8\frac12}$	2·800	${}^8D_{1\frac12}$
1·091	${}^2H_{5\frac12}$	1·414	${}^8G_{4\frac12}$	3·333	${}^6D_{\frac12}$
1·108	${}^4I_{6\frac12}$	1·429	${}^4D_{3\frac12}$	4·000	${}^8F_{\frac12}$

Terms of even multiplicity.

Fig. 6·12. Magnetic splitting factors.

strength of the magnetic field with a test coil and applying the Lorentz formula, but in practice one measures instead the splitting of a line of a known anomalous type.

An examination of all known Zeeman multiplets, or an application of general theory, shows that all can be considered as consisting of three groups of lines, a group of π components spreading across the position of the undisplaced line and a group of σ components on either side. Each group consists of a number of lines, whose separation is a constant, often written e. This value e must be determined, but as it may be obtained from the stronger components, the weaker components which may not have been measured do not hinder the work. Theory shows that $g_x - g_y = \pm e$. The only other value which needs to be measured is the distance between the two brightest σ components, a quantity commonly

g	Terms	g	Terms
−0·500	$^7G_1$	1·214	$^5H_6$
0/0	$^1S_0$, $^3P_0$, $^5D_0$, $^7F_0$	1·232	$^7I_7$
0·000	$^5F_1$, $^7H_2$	1·250	$^3F_4$, $^5F_3$, $^5I_8$
0·250	$^7I_3$	1·267	$^5G_5$
0·333	$^5G_2$	1·286	$^5H_7$, $^7H_6$
0·500	$^3D_1$, $^5H_3$	1·292	$^7I_8$
0·600	$^5I_4$	1·300	$^7G_4$
0·667	$^3F_2$	1·333	$^3D_3$, $^5G_6$, $^7I_9$
0·750	$^3G_3$, $^7H_3$, $^7I_4$	1·339	$^7H_7$
0·800	$^3H_4$	1·350	$^5F_4$
0·833	$^3I_5$, $^7G_2$	1·367	$^7G_5$
0·900	$^5H_4$, $^5I_5$	1·375	$^7H_8$
0·917	$^5G_3$	1·400	$^5F_5$
1·000	$^1P_1$, $^1D_2$, $^1F_3$, $^1G_4$, $^1H_5$, $^1I_6$, $^5F_2$, $^7I_5$	1·405	$^7G_6$
1·024	$^3I_6$	1·429	$^7G_7$
1·033	$^3H_5$	1·500	$^3P_{1,2}$, $^5D_{1,2,3,4}$, $^7F_{1,2,3,4,5,6}$
1·050	$^3G_4$, $^7H_4$	1·600	$^7D_5$
1·071	$^5I_6$	1·650	$^7D_4$
1·083	$^3F_3$	1·667	$^5P_3$
1·100	$^5H_5$	1·750	$^7P_4$, $^7D_3$
1·143	$^3H_7$, $^7I_6$	1·833	$^5P_2$
1·150	$^5G_4$	1·917	$^7P_3$
1·167	$^3D_2$, $^3H_6$, $^7G_3$	2·000	$^3S_1$, $^5S_2$, $^7S_3$, $^7D_2$
1·179	$^5I_7$	2·333	$^7P_2$
1·200	$^3G_5$, $^7H_5$	2·500	$^5P_1$
		3·000	$^7D_1$

Terms of odd multiplicity.

After Kiess and Meggers, *BSJ*, 1928, **1** 641.

written $2f$. Should there be a pair of equally bright lines as in Class III, then $2f$ is the distance from the centre of one pair to the centre of the other.

The number of π components will be always $(2J_x+1)$ provided we consider that the transition from $M_x=0$ to $M_y=0$ is not forbidden, but gives a line too weak to be observed. Thus if the number of π components is known, or if it be obtained by dividing the distance between two extreme bright components by e, then J_x can be calculated. To obtain J_y, note that if the intensities show the pattern to belong to Class I or II, $J_y=J_x+1$, while if it belongs to Class III $J_y=J_x$.

The values of J_x and J_y being known, g_x and g_y may be determined from e and f, but the precise form of the equations depends on the intensity class to which the Zeeman type belongs.

Class I. $J_x < J_y$ and $g_x < g_y$.

...	0	g_x	$2g_x$	...	$J_x g_x$ ↘	
...	0	g_y	$2g_y$	...		$J_y g_y$

In this figure the brightest σ component is shown by an arrow, so that

$$g_y J_y - g_x J_x = f.$$

Using the relations $$J_y = J_x + 1$$
and $$g_x = g_y - e,$$
we obtain $$g_y = f - J_x e.$$

Class II. $J_x < J_y$ and $g_x > g_y$.

A similar calculation shows that

$$\left.\begin{aligned} g_y &= f + J_x e \\ g_x &= g_y + e \end{aligned}\right\}.$$

Class III. $J_x = J_y$ and $g_x \neq g_y$.

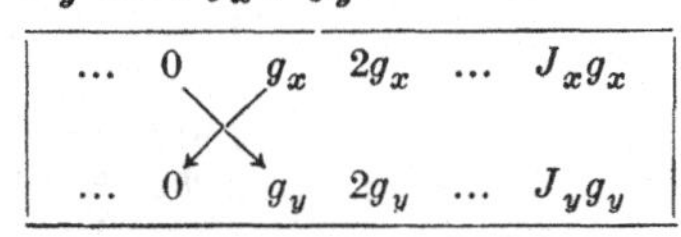

In this figure, drawn for terms of odd multiplicity, the brightest σ components are shown by arrows. Consequently, if $2f$ measures the mean of the distance between the two bright pairs of lines,

$$f = \tfrac{1}{2}(g_x + g_y).$$

And accordingly, if $g_x < g_y$,

$$\left.\begin{aligned} g_x &= f - e/2 \\ g_y &= f + e/2 \end{aligned}\right\}.$$

And these equations are easily shown to hold also for terms of even multiplicity.

Class IV. The anomalous Zeeman triplet may arise from (1) $e = 0$; when $g_x = g_y = f$, and J remains undetermined, (2) $J_x = 0$; when $g_x = \frac{0}{0}$, and $g_y = f$.

An example may help to clarify this discussion. The lines of 4163·64 A. and 4123·86 A. were observed in the spectrum of columbium, and their Zeeman effects measured by Jack.* They are respectively (–, *1·18*) 0·66, *1·46*, 2·24, and (*0·32*, 0·90) *0·47*, 1·02, 1·58, 2·18.

The first belongs to intensity type III, so that $J_x = J_y$. And

* Jack, *R. Irish Acad. Proc.* 1912, **30**A 42.

we find that $e=0{\cdot}79$, $f=1{\cdot}46$. Now the number of π components must be $\frac{1{\cdot}18\times 2}{e}+1$ from the measurements and $2J_x+1$ from theory, so that $J_x=1\frac{1}{2}$. Further,

$$g_x=f-e/2=1{\cdot}06,$$
$$g_y=f+e/2=1{\cdot}86.$$

Turning to Fig. 6·12, it appears that the only satisfactory identification is $^6F_{1\frac{1}{2}}$ and $^6D_{1\frac{1}{2}}$, two terms which have g values of 1·067 and 1·867 respectively.

For the line 4123·86 A. of columbium, we find $e=0{\cdot}57$ and $f=0{\cdot}47$. The variation of intensity shows the Zeeman type to belong to Class II so that $J_x=J_y-1$ and $g_x>g_y$. There are four π components, so that

$$(2J_x+1)=4$$

and $$J_x=1\tfrac{1}{2}, \quad J_y=2\tfrac{1}{2}.$$

This checks with the number of σ components which should be (J_x+J_y) either side of the centre, for in fact four were found.

Applying the formula $g_y=f+J_xe$,

we obtain $$g_y=1{\cdot}325,$$

and $$g_x=g_y+e=1{\cdot}875.$$

The only terms with approximately these g values and the correct values of J are $^6D_{1\frac{1}{2}}$ and $^6F_{2\frac{1}{2}}$; these have theoretical g values of 1·867 and 1·314 respectively.

The fact that this analysis has been made to depend only on the quantities e and f, on the relative intensities, and on the number of π components is of no little importance. For not only are the weak components difficult to measure, but if the magnetic field is strong enough to produce distortion e and f may still be calculated fairly accurately. Take e_σ and e_σ' as the mean values of e for the two groups of σ components, and e_π as the value of e for the group of π components, then the mean value

$$\bar{e}=\tfrac{1}{2}e_\pi+\tfrac{1}{4}(e_\sigma+e_\sigma')$$

is found to approximate closely to the ideal value. f is very little distorted by a strong field and needs no correction.

A good example of the application of this method to a distorted pattern may be taken from the 5218 A. line of copper, one line of a close doublet. The observed splitting in a field of $3{\cdot}7\,.\,10^4$ gauss is shown above the line in Fig. 6·13, the ideal splitting calculated after analysis is shown below.

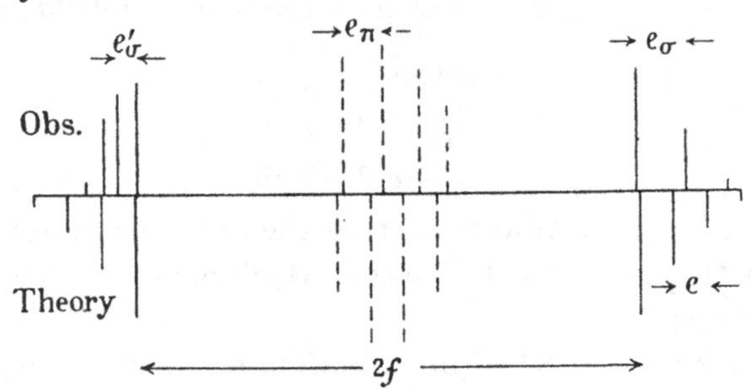

Fig. 6·13. Splitting of the 5218 A., $^2D_{2\frac{1}{2}} \rightarrow {}^2P_{1\frac{1}{2}}$, line of copper in a field of $3{\cdot}7 \,.\, 10^4$ gauss. Observed splitting above, ideal below. Observations after Back, *Hb. d. Expt. Phys.* 1929, **22** 107.

The observed Zeeman displacements are –, –1·350, –1·171, *–0·991*, (–0·240), (*–0·131*), (*0·012*), (0·169), *0·993*, 1·065, 1·123, 1·190. The separations of adjacent components are here very far from constant; indeed e_σ and e_σ', the mean values of e for the two groups of σ components, are 0·179 and 0·066, and so in a ratio of nearly 3 : 1. But taking the mean of these $\bar{e}_\sigma$ and combining it with the separation of the π components $e_\pi = 0{\cdot}136$, we obtain $\bar{e} = 0{\cdot}129$, which is not very far from the ideal value of 0·133. At any rate the error is not so great but that, proceeding with the analysis, we obtain

$$\left.\begin{array}{ll} g_x = 1{\cdot}314, & J_x = 1\frac{1}{2} \\ g_y = 1{\cdot}185, & J_y = 2\frac{1}{2} \end{array}\right\},$$

and so identify the terms without ambiguity as $^2P_{1\frac{1}{2}}$ and $^2D_{2\frac{1}{2}}$.

Thus even where the field required to produce a measurable splitting also produces distortion, the identity disc of the line is often legible.*

BIBLIOGRAPHY

The early work is thoroughly covered by Zeeman in *Magneto-optics*, 1914. Back and Landé in *Zeemaneffekt und Multiplettstruktur der Spektrallinien*, 1925, give a useful bibliography and some beautiful photographs, but their theoretical account has been superseded by Back's article in the *Handbuch der Experimental Physik*, 1929, **22**. This last is a very thorough study of both the Zeeman and Paschen-Back effects.

* For recent work on the use of unresolved patterns, see Shenstone and Blair, *PM*, 1929, **8** 765; Russell, *PR*, 1930, **36** 1590; Fisher, *PR*, 1933, **44** 724.

CHAPTER VII

PASCHEN-BACK EFFECT

1. Empirical

In the early years of the century, Voigt* examined several lithium lines and showed that they all split to normal triplets; yet the lithium spectrum is so similar to those of the other alkalis, that the principal series lines 6708 A. and 3233 A. must be unresolved doublets, and the diffuse lines 6104 A. and 4603 A. consist of three unresolved components. Again, the magnesium line 3838 A., arising as $3\,{}^3D \rightarrow 2\,{}^3P_2$, may be expected to behave as if the three components have been superposed and could be resolved with higher dispersion; thus the Zeeman type should be the result of superposing the patterns of the three mercury lines, 3650 A. $3\,{}^3D_3 \rightarrow 2\,{}^3P_2$, 3655 A. $3\,{}^3D_2 \rightarrow 2\,{}^3P_2$, and 3663 A. $3\,{}^3D_1 \rightarrow 2\,{}^3P_2$; though in fact the 3838 A. line exhibits a simpler pattern characteristic of the transition ${}^3S_1 \rightarrow {}^3P_2$. Most physicists accordingly catalogued all these lines as exceptions to Preston's rule,* but in 1912 Paschen and Back† advanced another explanation.

They argued that in a weak field each line of a multiplet splits into a number of Zeeman components, whose displacements are proportional to the strength of the field. If this simple law does not break down when the strength of the field is increased, the components of the two lines must cross, and this will occur with quite weak fields if the doublet interval is small. At first sight it seems easy to watch the changes in the sodium doublet as the field is increased; but the strongest field available is limited to 45,000 gauss, and this produces a Lorentz interval of only $2{\cdot}1$ cm.$^{-1}$, or $0{\cdot}76$ A. in a line at 6000 A.; thus the Zeeman components of the D lines never meet, and one has to assume that, like the doublets of lithium, the two patterns would merge as the field is increased, and that in a very strong field there is left just

* Zeeman, *Magneto-optics*, 1913, 165.

† Paschen and Back, *AP*, 1912, **39** 897; 1913, **40** 960.

a simple Zeeman triplet in which the two σ components are displaced by

$$o_m = \frac{eH}{4\pi m_e c^2} \text{ cm.}^{-1}$$

Thus the condition for the appearance of the Paschen-Back triplet is that the magnetic displacement shall be much greater than the intervals of the undisturbed multiplet, or symbolically, that $o_m \gg \Delta\nu$; and conversely the Zeeman splitting of a line will be regular only when $o_m \ll \Delta\nu$. The pattern produced therefore depends on the ratio $o_m/\Delta\nu$, and not at all on the absolute value of the field; and so in the further discussion, a 'weak' field will be taken to mean one in which this fraction is small, and a 'strong' field one in which it is large; the words are not to be thought to refer to the absolute strength of the field.

Clearly the best hope of confirming this hypothesis lies in observing the narrowest known multiplet; Paschen and Back* accordingly photographed a principal triplet of oxygen, whose lines are of wave-length 3947·438, 3947·626 and 3947·731 A. This fulfilled expectations; in a field of 6000 gauss the lines are so wide that they are no longer resolved; thereafter the resolving power is insufficient to distinguish the components predicted by theory, but when the field reaches 32,000 gauss a simple triplet is formed, in which the central component is quite sharp, though the side components are still somewhat diffuse; in a later section theory will be seen to predict this contrast.

In order to make this picture more precise, consider the changes which take place in the $D_1 D_2$ lines of sodium, or any other principal doublet, though it is well to be clear that in describing the detailed changes we rely as much on theory as experiment. In Fig. 7·1† the weak field components are shown on the left, and the strong field components on the right, the centre of the strong field pattern coinciding with the centroid of the undisturbed doublet. The directions in which the various components move are here clearly shown, but the changes in intensity are only

* Paschen and Back, *AP*, 1912, **39** 897; 1913, **40** 960. The photographs are reproduced in these papers, and in Back and Landé, *Zeemaneffekt und Multiplettstruktur*. 1925.

† Voigt, *AP*, 1913, **42** 210.

adumbrated; as, for instance, by the fact that there are finally no components with a displacement of $2o_m$, so that components which move towards this position must fade. For greater detail

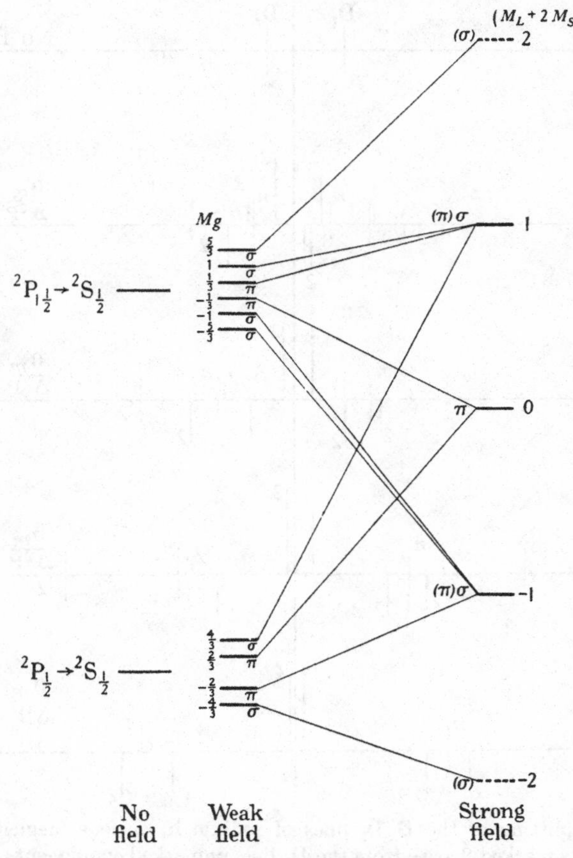

Fig. 7·1. Diagrammatic sketch of the changes occurring in a principal doublet as the field is increased; where π or σ is enclosed in brackets, this component fades in a strong field.

turn to Fig. 7·2,* in which the heights of the lines represent the intensities, and the positions of the various components are given precisely for various field strengths. The components labelled 2 derive from the D_2 line, those not labelled from the D_1 line; the

* Voigt, *AP*, 1913, 42 210.

dotted lines labelled n show the size of the Lorentz unit, for they mark the positions of the σ components of an imaginary singlet line initially at the centroid of the doublet. This second figure is

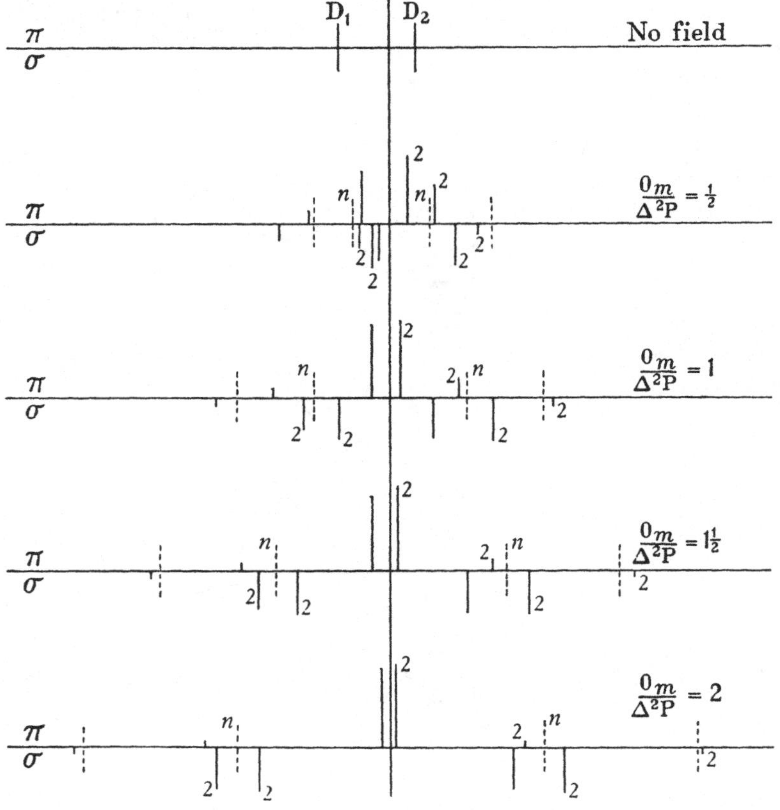

Fig. 7·2. Splitting of the D_1 D_2 lines of sodium in various magnetic fields. Components marked 2 arise from the D_2 line, unmarked components from D_1. The heights indicate intensities, and the dotted lines show the normal and twice normal intervals. After Voigt, *AP*, 1913, **42** 228.

taken from a theoretical paper by Voigt, but evidence will be adduced later to show that this is what would be observed under favourable conditions.

A last figure, Fig. 7·3 *a*, shows a third way of representing this same transition from weak to strong;* the dotted lines represent

* Darwin, K. *PRS*, 1928, **118** 264.

σ components, the full lines π components, while the breadth of each line is proportional to the intensity. In order to reduce the figure to a reasonable compass, however, the horizontal scale is

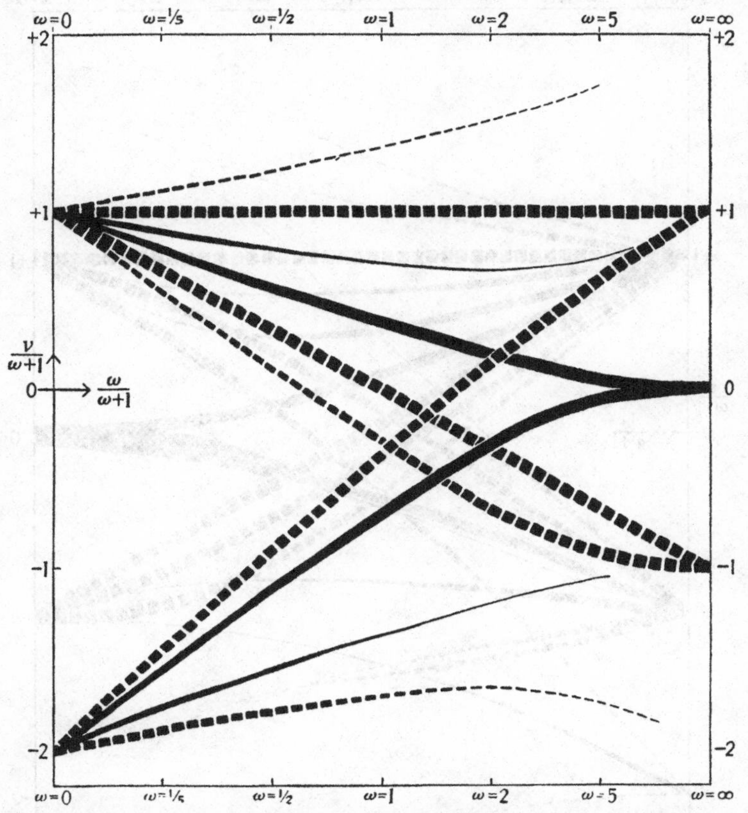

Fig. 7·3 *a*. Splitting of a sharp or principal doublet. The coordinates are distorted so as to include the strong fields, while showing detail in the weak. The abscissa run from zero to infinite field in the scale $\omega/\omega+1$, where ω is proportional to the field strength H. The ordinates measure $\nu/\omega+1$, where ν is the displacement from the centroid of the multiplet. The intensities are shown roughly by the breadths of the lines, the continuous lines being π and the broken σ components. After Darwin, K. *PRS*, 1928, **118** 264.

not simply the field strength, but a function of this, $\omega/(\omega+1)$, which tends to unity when the field tends to infinity; ω is here the Larmor frequency, a quantity proportional to the Lorentz interval o_m. Again, the ordinates are not displacements of the lines from the centroid, written by Mrs Darwin ν, but $\nu/(\omega+1)$. This method of representation gives a bird's-eye view of the whole

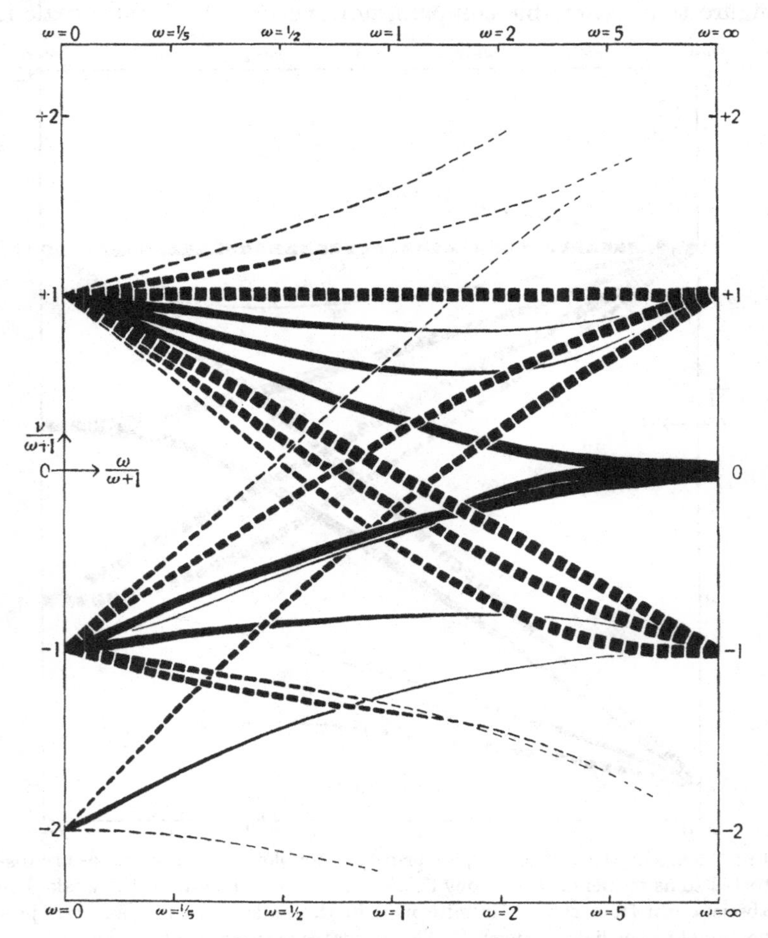

Fig. 7·3 *b*. Splitting of a sharp or principal triplet. The symbolism is the same as in the preceding figure.

change, but it is more difficult to compare numerical data with this figure than with Voigt's.

Generalising these results, we see that in any multiplet, as the field is increased, the simple Zeeman laws fail and the components flow as if they attract and repel one another; this change is far from simple, while some components grow brighter, others fade and finally disappear. Whatever the multiplet, however, there emerges in a sufficiently strong field the simple Zeeman triplet with a π component in the centre, a σ component on either side and the normal interval.

2. The vector model*

The transition from Zeeman to Paschen-Back effect is beautifully explained by the vector model; a weak field does not break the coupling of the orbital and spin vectors, but merely makes them precess round their resultant **J**, while **J** in turn precesses round the magnetic axis **H**; a strong field, on the other hand, breaks the coupling of **L** and **S**, and makes these two vectors precess round **H** independently. A convenient symbolism brackets the more firmly linked vectors, so that the Zeeman coupling is written {(**LS**) **H**}, while the coupling producing the Paschen-Back effect is {(**LH**) (**SH**)}. The more firmly two vectors are linked in the model, the greater the displacement of the terms due to the linking.

Moreover, the vector model indicates some such transition as is actually observed; for the coupling of **H** with **L**, (**HL**), will presumably increase with H, since in the singlet system, where $S=0$, (**HL**) increases uniformly with H throughout; and so, if the field is increased without limit, (**HL**) must finally become more important than (**LS**), since the latter is independent of the field.

Whereas in the weak field of the Zeeman effect the projection of **J** on **H** had to be quantised, in a strong field it is the projections of **L** and **S**, whose values are restricted. The appropriate quantum numbers are commonly written M_L and M_S, and the values they may assume are naturally restricted by the conditions

$$L \geqslant M_L \geqslant -L,$$
$$S \geqslant M_S \geqslant -S.$$

And whereas the Zeeman displacement is given by

$$\Delta E/ch = Mgo_m,$$

* Goudsmit, *PR*, 1929, **31** 946.

the Paschen-Back displacement is

$$\Delta E/ch = (M_L + 2M_S)\,o_m. \qquad \text{......(7·1)}$$

But while the former is measured from the position of the undisturbed L_J term, the latter is the displacement from the centroid of the multiplet. The factor 2 which occurs in this equation is due to the double magnetism of the electron, a point fully discussed in the preceding chapter.

Linking this up with equation (4·1), the energy of a level in a magnetic field may be written

$$E/ch = \nu_G + \Gamma_{w.} + Mgo_m \qquad \text{in a weak field,} \qquad \text{......(7·2)}$$

$$E/ch = \nu_G + \Gamma_{st.} + (M_L + 2M_S)\,o_m \qquad \text{in a strong field.} \qquad \text{......(7·3)}$$

The second term in these expressions results from the interaction of **L** and **S**, so that the displacement Γ is always equal to $LS\cos(\mathbf{LS})$; but while in a weak field $\cos(\mathbf{LS})$ is constant and can be evaluated simply in terms of L, S and J, in a strong field the value of the cosine varies and its mean value is required. **L** and **S** are both precessing round **H**, so

$$LS\,\overline{\cos(\mathbf{LS})} = LS\cos(\mathbf{LH})\cos(\mathbf{SH})$$
$$= M_L M_S. \qquad \text{......(7·4)}$$

When the terms of equation (7·3) are rearranged in order of decreasing magnitude, it becomes

$$E/ch = \nu_G + (M_L + 2M_S)\,o_m + M_L M_S A. \qquad \text{......(7·5)}$$

Before applying this theory to the D lines of sodium, new selection and polarisation rules have to be found. Experiment

Term	M_L	M_S	M_L+2M_S	$M_L M_S$	$\Delta E/ch$
^{2}S	0	$\frac{1}{2}$	1	0	o_m
		$-\frac{1}{2}$	-1	0	$-o_m$
^{2}P	1	$\frac{1}{2}$	2	$\frac{1}{2}$	$2o_m+\frac{1}{2}A$
		$-\frac{1}{2}$	0	$-\frac{1}{2}$	$-\frac{1}{2}A$
	0	$\frac{1}{2}$	1	0	o_m
		$-\frac{1}{2}$	-1	0	$-o_m$
	-1	$\frac{1}{2}$	0	$-\frac{1}{2}$	$-\frac{1}{2}A$
		$-\frac{1}{2}$	-2	$\frac{1}{2}$	$-2o_m+\frac{1}{2}A$

Fig. 7·4. Splitting of the ^{2}S and ^{2}P terms in a strong magnetic field.

shows that the facts can be explained if the only transitions permitted are those having $\Delta M_L = 0$ or ± 1 and $\Delta M_S = 0$, while in the polarisation rule of Bohr and Rubinowicz M_L is to be substituted for M. A jump having $\Delta M_L = 0$ corresponds to a π component, while $\Delta M_L = \pm 1$ gives a σ component.

Figs. 7·4 and 7·5 show how these rules work out in the D lines;

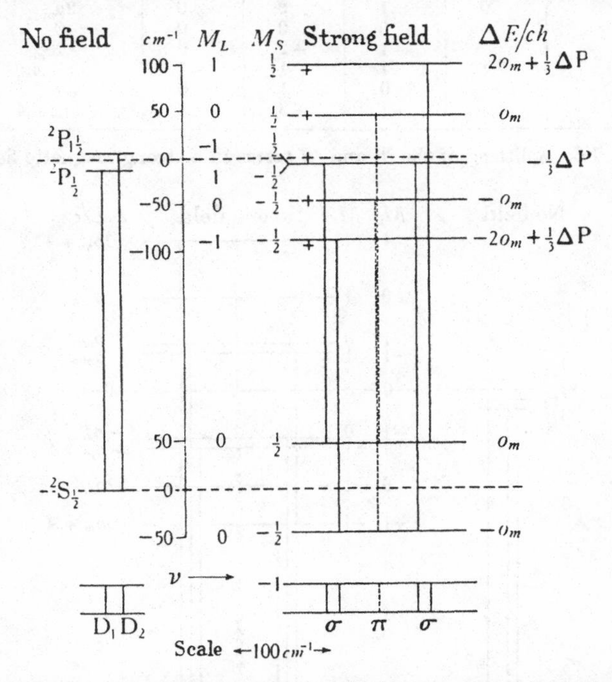

Fig. 7·5. Transitions producing a principal doublet in a strong field; the pattern resulting is a normal Zeeman triplet. The scale inserted is correct for the D_1 D_2 doublet of sodium, in a field of 1,000,000 gauss.

while Figs. 7·6 and 7·7 show similar calculations for a ^3S–^3P combination. In Figs. 7·4 and 7·6 displacements have been written in terms of the interval quotient A, which is $\frac{2}{3}$ of the doublet interval ΔP. The crosses in Fig. 7·5 show the positions that the P levels would occupy, if the term due to the interaction of the **L** and **S** vectors was dropped; the σ components would then exhibit no fine structure. The experimental difficulties however are so great that it seems impossible to decide whether the

Term	M_L	M_S	M_L+2M_S	$M_L M_S$	$\Delta E/ch$
^{3}S	0	1	2	0	$2o_m$
		0	0	0	0
		−1	−2	0	$-2o_m$
^{3}P	1	1	3	1	$3o_m+A$
		0	1	0	o_m
		−1	−1	−1	$-o_m-A$
	0	1	2	0	$2o_m$
		0	0	0	0
		−1	−2	0	$-2o_m$
	−1	1	1	−1	o_m-A
		0	−1	0	$-o_m$
		−1	−3	1	$-3o_m+A$

Fig. 7·6. Splitting of the ^{3}S and ^{3}P terms in a strong magnetic field.

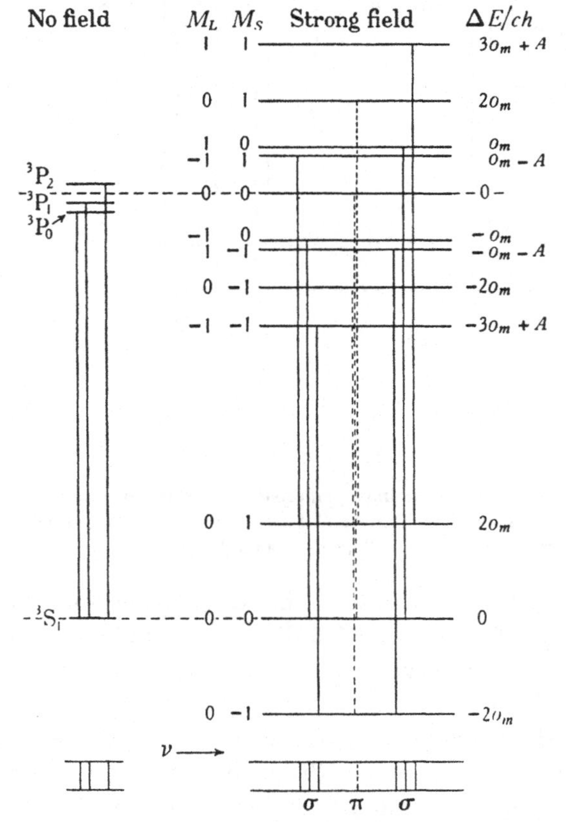

Fig. 7·7. Transitions producing a principal triplet in a strong magnetic field. The pattern is a normal triplet.

three lines of the Paschen-Back triplet are really simple or not; theory suggests that the components should have a structure of the order of magnitude of the original multiplet, but experimental confirmation is difficult, because one must use a narrow multiplet; these occur only in light atoms or high series terms, and in both the lines are far from sharp. That the σ components of the oxygen triplet 3947 A. are more diffuse than the π components is apparent, but one cannot place great faith in the observation when the photographs themselves are so small.

Fortunately the predictions of the vector model can be supported by other arguments; the quantum mechanics gives the same results in a strong field, and the quantum mechanics has been confirmed by measurements in intermediate fields.

3. Partial Paschen-Back effect

The multiplets so far considered have arisen from the combination of a doublet or a triplet term with an S term. When, however, two multiplets combine, a field which is strong to one may still be weak to the other, for the numerical strength of the field is no criterion of whether it is 'strong' or 'weak' in the technical sense. When this happens the pattern is said to be an example of the 'partial Paschen-Back' effect, to which Sommerfeld* first drew attention.

The diffuse series of sodium and magnesium afford excellent examples of this effect, for the interval separating the D levels is less than 3/10 cm.$^{-1}$, while the $2\,{}^2$P interval of sodium is 17 cm.$^{-1}$; and so while a field of 40,000 gauss giving a Lorentz unit of 1·9 cm.$^{-1}$ is strong to the D levels, the P levels still show the simple Zeeman splitting.

Back† has measured some of these lines, and his results are shown in Fig. 7·8. The Runge number of the type is always also the denominator of the splitting factor of the large interval term, a fact which gave rise to the explanation that in the partial Paschen-Back effect the small interval term behaves as a singlet whatever its true multiplicity. But as the Runge number of

* Sommerfeld, *AP*, 1920, **63** 221.
† Back, *AP*, 1923, **70** 370; *ZP*, 1925, **33** 588.

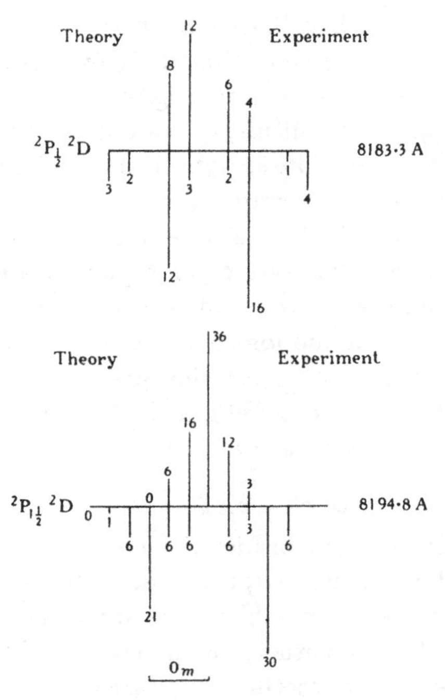

Fig. 7·8. Partial Paschen-Back effect of a diffuse doublet of sodium. Back's observations (*ZP*, 1925, **33** 579) are here compared with Mensing's calculations (*ZP*, 1926, **39** 24).

Term							
Term	M		$-1\frac{1}{2}$	$-\frac{1}{2}$	$\frac{1}{2}$	$1\frac{1}{2}$	
$^2P_{1\frac{1}{2}}$	Mg		-2	$-\frac{2}{3}$	$\frac{2}{3}$	2	
$^2D_{1\frac{1}{2}}$	M_L+2M_S		-2	-1	0	1	
$\Delta E/ch o_m$	$\dfrac{-3\ -2\ -1\ (0)\ (1)\ (2)\ (3)\ 4\ 5\ 6}{3}$						
Term	M	$-2\frac{1}{2}$	$-1\frac{1}{2}$	$-\frac{1}{2}$	$\frac{1}{2}$	$1\frac{1}{2}$	$2\frac{1}{2}$
$^2P_{1\frac{1}{2}}$	Mg		-2	$-\frac{2}{3}$	$\frac{2}{3}$	2	
$^2D_{2\frac{1}{2}}$	M_L+2M_S	-3	-1	0	1	2	3
$\Delta E/ch o_m$	$\dfrac{-6\ -5\ -4\ -3\ (-3)\ (-2)\ (-1)\ (0)\ 0\ 1\ 2\ 3}{3}$						

Fig. 7·9. Calculation of the partial Paschen-Back pattern of a $^2P_{1\frac{1}{2}} \rightarrow {}^2D$ line. The two components are treated separately, but one assumes that the ^2D interval is small, and that the patterns fuse.

terms subjected to a strong field is unity, these facts admit of another interpretation; the magnetic components of the small interval term may be assumed displaced by $(M_L + 2M_S)$, just as they are if the interval is large.

The second alternative, though more complex, is theoretically far preferable, and if the selection and polarisation rules of the weak field are assumed still valid, it leads to results which agree satisfactorily with experiment. Thus Fig. 7·9 shows the calculation for the $^2P_{1\frac{1}{2}}\,^2D_{1\frac{1}{2}}$ and $^2P_{1\frac{1}{2}}\,^2D_{2\frac{1}{2}}$ lines arranged separately; combined, these predict the pattern of the $^2P_{1\frac{1}{2}}\,^2D$ in a magnetic field as (0) (1) (2) (3) 1 2 3 4 5 6/3, whereas the observed type is (0) (1) (2) 1 2 3 4/3. The two agree, if it is assumed that three of the components predicted are too weak to observe; and this agrees with some intensity calculations made by Mensing* on the basis of the quantum mechanics (Fig. 7·8).

More recently Van Geel has shown that the magnesium triplets also provide satisfactory confirmation of theory.†

4. Intermediate fields

The vector model is able to deal only with fields which are either strong or weak, but Voigt‡ early developed a method capable of predicting the structure of the $D_1 D_2$ lines in any field (Fig. 7·2); more recently the quantum mechanics, successfully applied to the general multiplet, has confirmed Voigt's work.§

Calculations based on the quantum mechanics support the predictions of the vector model in weak and strong fields, but it is important to show that they are satisfactory where they are most vulnerable, namely in intermediate fields. The only accurate measurements available for this purpose seem to be those of Back,|| who observed three multiplets, a $^2P\,^2D$ multiplet of Cu near 5220 A., a $^3P\,^3S$ multiplet of Be at 3322 A., and a $^3P\,^3S$ multiplet of Mg near 5167 A. Recently Green¶ has calculated what

* Mensing, *ZP*, 1926, **39** 24. † See chapter XVII on Intensity.

‡ Voigt, *AP*, 1913, **42** 210. Cf. Sommerfeld, *Atomic structure and spectral lines*, 1923, 400f.

§ Heisenberg and Jordan, *ZP*, 1926, **37** 263; Darwin, C. G. *PRS*, 1927, **115** 1.

|| Back, *AP*, 1923, **70** 350. ¶ Green, *PR*, 1930, **36** 157.

splitting the quantum mechanics predicts for these three lines, and has shown that the agreement is good.

Here we need examine only one of these multiplets, and choose the $^2P\,^2D$ of Cu I. In this multiplet $A = 6{\cdot}90/5 = 1{\cdot}38$ cm.$^{-1}$ for the

Position of undisturbed line	Position in magnetic field			Intensities		
	Obs.	Calc. by wave mechanics	Calc. by old quantum theory	Obs.	Calc. by wave mechanics	Calc. by old quantum theory
5218·200 $^2D_{2\frac{1}{2}} \rightarrow {}^2P_{1\frac{1}{2}}$	18·764	18·835	18·863	1	1	1
	·735	·770	·800	6	3	3
	·704	·715	·737	7	5	5
	·670	·673	·673	8	8	8
	·279	·276	·295	7	7	7
	·205	·205	·232	10	10	10
	·137	·137	·168	10	9·5	10
	·085	·090	·105	7	6	7
	17·729	17·727	17·727	8	8	8
	·644	·647	·663	7	4·5	5
	·560	·573	·660	1	2	3
	—	·508	·537	0	0·5	1
5220·080 $^2D_{1\frac{1}{2}} \rightarrow {}^2P_{1\frac{1}{2}}$	20·857	20·864	20·837	2	1	3
	·600	·615	·585	5	4	4
	·433	·474	·459	8	4	10
	·282	·354	·332	4	1	3
	·148	·235	·206	1	1	1
	19·959	19·985	19·954	1	2	1
	·845	·846	·828	4	3	3
	·727	·726	·701	10	10	10
	·608	·604	·575	5	4	4
	·367	·356	·323	3	3	3
5153·26 $^2D_{1\frac{1}{2}} \rightarrow {}^2P_{\frac{1}{2}}$	53·652	53·652	53·660	7	7·5	7·5
	—	·599	·598	0	2·5	2·5
	·292	·291	·291	10	10	10
	·228	·229	·229	10	10	10
	52·909	52·921	52·922	4	2·5	2·5
	·850	·850	·862	9	7·5	7·5

Fig. 7·10. Splitting of a diffuse doublet of copper in an intermediate field. Back's measurements are here compared with the predictions of the old quantum theory and the wave mechanics; the interval factors of the ^2D term is $6{\cdot}90/5 = 1{\cdot}38$ cm.$^{-1}$, and of the ^2P term $248{\cdot}3/3 = 82{\cdot}8$. Back used a field of 37,500 gauss so that $o_m = 1{\cdot}76$ cm.$^{-1}$

^2D levels, and $248{\cdot}3/3 = 82{\cdot}8$ cm.$^{-1}$ for the ^2P levels, while $o_m = 1{\cdot}76$ cm.$^{-1}$ A study of the figures obtained (Fig. 7·10) shows that except for one component, 5220·433, the wave mechanics gives much better agreement with experiment than the older theory; all other exceptions to this statement are faint lines which cannot be accurately measured; the intensities are also distorted as the newer theory suggests.

That so important a theory should rest on a single set of measurements appears unfortunate; but in fact the foundations are somewhat stronger than this, for the theory developed for gross multiplets applies also to hyperfine multiplets; several measurements of the latter are available and these are always in satisfactory agreement with theory.

5. Matching weak and strong terms

That the Zeeman components of a line are sharp, whatever the strength of the field, shows that the energy states must always be determined by quantum conditions. If the field is slowly made stronger, each of these states will retain its identity, so that a Zeeman term will change slowly into a Paschen-Back term. This has two consequences; first, the number of terms in a strong field must be the same as the number in a weak; and second, a state characterised by the weak field quantum numbers L, S, J and M will change to one characterised by the strong field numbers L, S, M_L and M_S, so that we are forced to ask what values of J and M correspond to given values of M_L and M_S.

That the quantum numbers do in fact yield the same number of orbits in weak and strong fields is easily demonstrated. Consider, for example, a ^2L term; in a weak field the permitted values of J are $(L+\frac{1}{2})$ and $(L-\frac{1}{2})$, and as each of these gives rise to $(2J+1)$ components there are $(4L+2)$ components in all; in a strong field, on the other hand, M_L may assume $(2L+1)$ values, and M_S $(2S+1)$ values, so that the number of components is $(2L+1)(2S+1)$; and this general expression assumes the value $(4L+2)$ for a doublet term having $S=\frac{1}{2}$. The summing of the weak field components of multiplicity greater than 2 is a little more trouble, but the result is always $(2L+1)(2S+1)$.

There remains the second question; what values of J and M are to be matched to given values of M_L and M_S? M is easily determined, for it is simply the sum of M_L and M_S; J however is more troublesome. The theoretical answer was worked out by Sommerfeld and Pauli* with the help of the correspondence

* Sommerfeld, *ZP*, 1922, **8** 257; Pauli, *ZP*, 1923, **20** 371.

principle, and confirmed by Heisenberg and Jordan* with the quantum mechanics. It is extremely simple in form, for it states only that levels having the same value of M do not cross in the transition from a weak to a strong field. But as this solution is not easily worked out in detail, we shall first describe an empirical solution due to Breit,† and then show that Heisenberg's solution leads to the same result.

The components of an ${}^SL$ term are conveniently fitted into the cells of a rectangle, which is $(2L+1)$ in length and $(2S+1)$ in depth. Each cell can then be labelled with a unique value of M_L and M_S, the values of M_L being made to increase from left to right, while the values of M_S increase from bottom to top. In each cell inscribe values of M as in Fig. 7·11, values of (M_L+2M_S) as in Fig. 7·12, or values of $M_L M_S$ as in Fig. 7·13; and the only question which remains is 'How are we to allot J?' Breit answered this question by dividing the rectangle into alleys each one square in width with a right-angled bend in the top left-hand corner. The first alley contains eight cells with values of M ranging from $-3\frac{1}{2}$ to $3\frac{1}{2}$, and these compose the ${}^4\mathrm{D}_{3\frac{1}{2}}$ term; the second alley has six cells, which compose the ${}^4\mathrm{D}_{2\frac{1}{2}}$ term; and similarly the other two alleys contain the components of the ${}^4\mathrm{D}_{1\frac{1}{2}}$ and ${}^4\mathrm{D}_{\frac{1}{2}}$ terms. Thus if a component is first defined by values of M_L and M_S, so that the position in a strong field is known, Breit's system of alleys fixes J and M, and hence the position in a weak field.

This method may be applied to any ${}^SL$ term, provided it is erect. Should it be inverted the values of M_L and M_S may be arranged as before, but the L-shaped alleys must be drawn with their angles in the bottom right-hand corner. Figs. 7·14 and 7·15 illustrate this for an inverted ${}^5\mathrm{P}$ term.

Breit's alleys contain the whole solution of the matching problem, but certain other ways of arranging the results are worth notice. Thus the transition from a weak to a strong field may be illustrated by writing the displacements in the weak field, Mg, on the left, and those in the strong field, (M_L+2M_S), on

* Heisenberg and Jordan, *ZP*, 1926, **37** 263.

† Breit, *PR*, 1926, **28** 334; Russell, *PR*, 1927, **29** 783.

M_S \ M_L	-2	-1	0	1	2
$1\frac{1}{2}$	$-\frac{1}{2}$	$\frac{1}{2}$	$1\frac{1}{2}$	$2\frac{1}{2}$	$3\frac{1}{2}$
$\frac{1}{2}$	$-1\frac{1}{2}$	$-\frac{1}{2}$	$\frac{1}{2}$	$1\frac{1}{2}$	$2\frac{1}{2}$
$-\frac{1}{2}$	$-2\frac{1}{2}$	$-1\frac{1}{2}$	$-\frac{1}{2}$	$\frac{1}{2}$	$1\frac{1}{2}$
$-1\frac{1}{2}$	$-3\frac{1}{2}$	$-2\frac{1}{2}$	$-1\frac{1}{2}$	$-\frac{1}{2}$	$\frac{1}{2}$
	$^4D_{3\frac{1}{2}}$	$^4D_{2\frac{1}{2}}$	$^4D_{1\frac{1}{2}}$	$^4D_{\frac{1}{2}}$	

Fig. 7·11. The matching of M_L and M_S with J in an erect ^4D term. The values inserted in the body of the figure are M_L+M_S, or what is equivalent M_J.

M_S \ M_L	-2	-1	0	1	2
$1\frac{1}{2}$	1	2	3	4	5
$\frac{1}{2}$	-1	0	1	2	3
$-\frac{1}{2}$	-3	-2	-1	0	1
$-1\frac{1}{2}$	-5	-4	-3	-2	-1
	$^4D_{3\frac{1}{2}}$	$^4D_{2\frac{1}{2}}$	$^4D_{1\frac{1}{2}}$	$^4D_{\frac{1}{2}}$	

Fig. 7·12. Values of (M_L+2M_S) in an erect ^4D term; the value of J corresponding to a given value of (M_L+2M_S) is found at the end of the alley, the corresponding value of M in the same cell of the previous figure.

M_S \ M_L	-2	-1	0	1	2
$1\frac{1}{2}$	-3	$-1\frac{1}{2}$	0	$1\frac{1}{2}$	3
$\frac{1}{2}$	-1	$-\frac{1}{2}$	0	$\frac{1}{2}$	1
$-\frac{1}{2}$	1	$\frac{1}{2}$	0	$-\frac{1}{2}$	-1
$-1\frac{1}{2}$	3	$1\frac{1}{2}$	0	$-1\frac{1}{2}$	-3
	$^4D_{3\frac{1}{2}}$	$^4D_{2\frac{1}{2}}$	$^4D_{1\frac{1}{2}}$	$^4D_{\frac{1}{2}}$	

Fig. 7·13. Values of $M_L M_S$ in an erect ^4D term; the values of J and M may be found as in the previous figure.

$^5P_1$	$^5P_2$	$^5P_3$	M_S
1	2	3	2
0	1	2	1
-1	0	1	0
-2	-1	0	-1
-3	-2	-1	-2
-1	0	1	M_L

Fig. 7·14. The matching of M_L and M_S with J in an inverted ^{5}P term. The values of M are inserted.

$^5P_1$	$^5P_2$	$^5P_3$	M_S
3	4	5	2
1	2	3	1
-1	0	1	0
-3	-2	-1	-1
-5	-4	-3	-2
-1	0	1	M_L

Fig. 7·15. Values of (M_L+2M_S) in an inverted ^{5}P term.

$^5P_1$	$^5P_2$	$^5P_3$	M_S
$\frac{5}{2} \rightarrow 3$	$\frac{11}{3} \rightarrow 4$	$5 \rightarrow 5$	2
$0 \rightarrow 1$	$\frac{11}{6} \rightarrow 2$	$\frac{10}{3} \rightarrow 3$	1
$-\frac{5}{2} \rightarrow -1$	$0 \rightarrow 0$	$\frac{5}{3} \rightarrow 1$	0
$-\frac{11}{3} \rightarrow -3$	$-\frac{11}{6} \rightarrow -2$	$0 \rightarrow -1$	-1
$-5 \rightarrow -5$	$-\frac{10}{3} \rightarrow -4$	$-\frac{5}{3} \rightarrow -3$	-2
-1	0	1	M_L

Fig. 7·16. The transition from the weak field displacement Mg to the strong field (M_L+2M_S) is shown for each component of an inverted ^{5}P term. The components are arranged by their M_L and M_S values.

Term \ M	-3	-2	-1	0	1	2	3	M / Term
$^5P_1$	—	—	$-\frac{5}{2} \rightarrow -1$	$0 \rightarrow 1$	$\frac{5}{2} \rightarrow 3$	—	—	$^5P_1$
$^5P_2$	—	$-\frac{11}{3} \rightarrow -3$	$-\frac{11}{6} \rightarrow -2$	$0 \rightarrow 0$	$\frac{11}{6} \rightarrow 2$	$\frac{11}{3} \rightarrow 4$	—	$^5P_2$
$^5P_3$	$-5 \rightarrow -5$	$-\frac{10}{3} \rightarrow -4$	$-\frac{5}{3} \rightarrow -3$	$0 \rightarrow -1$	$\frac{5}{3} \rightarrow 1$	$\frac{10}{3} \rightarrow 3$	$5 \rightarrow 5$	$^5P_3$

Fig. 7·17. $Mg \rightarrow (M_L+2M_S)$ as in the previous figure, but the components are here arranged by their J and M values.

the right of the cell (Fig. 7·16). Back and Lande used a table in this form, but arranged the components according to their J and M values, as in Fig. 7·17.

And how does Heisenberg's pronouncement, that terms having the same value of M do not cross, fit in with this empirical solution? To obtain an answer, consider those components of an erect ^{4}D term which have M values of $3\frac{1}{2}$ and $\frac{1}{2}$, for these sufficiently illustrate the argument. There is only one component in a weak or a strong field which has $M = 3\frac{1}{2}$, so that these must be matched on any theory. When $M = \frac{1}{2}$, however, four components having $J = 3\frac{1}{2}$, $2\frac{1}{2}$, $1\frac{1}{2}$ and $\frac{1}{2}$ have to be matched with strong components having $(M_L + 2M_S) = 2$, 1, 0 and -1; since the term is erect, the greatest J will lie highest in the energy scale in the absence of a magnetic field, and must be matched with $M_L + 2M_S = 2$, which lies highest in a strong field; and similarly, the component lying second highest in a weak field, namely that having $J = 2\frac{1}{2}$, must also lie second highest in the strong field and so have $M_L + 2M_S = 1$. These results, as well as those of the other two components, are clearly those obtained in Figs. 7·11 and 7·12 from Breit's alleys.

Moreover, the agreement is equally satisfactory when the term is inverted. Those components of the ^{5}P term which have $M = 1$ have $J = 3$, 2 or 1 and $(M_L + 2M_S) = 3$, 2 or 1; but when the term is inverted the $J = 1$ component lies highest, and therefore has to be matched with the strong component having $M_L + 2M_S = 3$. This is exactly what was deduced in Figs. 7·14 and 7·15.

Though Breit's alleys are so satisfactory for erect and inverted terms, they cannot be applied to terms in which the J sequence is irregular, terms which are commonly said to be partially inverted. Such a term is the $^3P_2$ ground term of tellurium, in which the $J = 0$ component lies between those having J values of 1 and 2. Heisenberg's rule does predict exactly the transition which should occur, but it seems that this has not yet been tested by experiment.

6. Invariance of the *g* sum

Among a group of terms all of which have the same value of L, S and M, the sum of the magnetic displacements is the same in a weak as in a strong field.* The sum of the displacements in a weak field is simply $\sum_J Mg$ summed over all values of J, and since M is constant, this reduces to $M \sum_J g$. In a strong field the sum is $\Sigma(M_L + 2M_S)$ summed over all values of M_L and M_S, which satisfy the condition that $(M_L + M_S) = M$. Thus

$$\sum_J Mg = \sum_{M_S} (M_L + 2M_S),$$

when L, S and M are constant.

This rule is easily tested for the ^{4}D term by summing the vertical columns of Fig. 7·18, where it appears that for $M = \frac{1}{2}, 1\frac{1}{2}, 2\frac{1}{2}$ and $3\frac{1}{2}$, the sums of the displacements are 2, 6, 7 and 5 respectively.

Term \ M	$-3\frac{1}{2}$	$-2\frac{1}{2}$	$-1\frac{1}{2}$	$-\frac{1}{2}$	$\frac{1}{2}$	$1\frac{1}{2}$	$2\frac{1}{2}$	$3\frac{1}{2}$	M / Term
$^4D_{\frac{1}{2}}$				0	0				$^4D_{\frac{1}{2}}$
$^4D_{1\frac{1}{2}}$			$-\frac{9}{5}$	$-\frac{3}{5}$	$\frac{3}{5}$	$\frac{9}{5}$			$^4D_{1\frac{1}{2}}$
$^4D_{2\frac{1}{2}}$		$-\frac{24}{7}$	$-\frac{72}{35}$	$-\frac{24}{35}$	$\frac{24}{35}$	$\frac{72}{35}$	$\frac{24}{7}$		$^4D_{2\frac{1}{2}}$
$^4D_{3\frac{1}{2}}$	-5	$-\frac{25}{7}$	$-\frac{15}{7}$	$-\frac{5}{7}$	$\frac{5}{7}$	$\frac{15}{7}$	$\frac{25}{7}$	5	$^4D_{3\frac{1}{2}}$
ΣM_g	-5	-7	-6	-2	2	6	7	5	ΣM_g

Term \ M	$-3\frac{1}{2}$	$-2\frac{1}{2}$	$-1\frac{1}{2}$	$-\frac{1}{2}$	$\frac{1}{2}$	$1\frac{1}{2}$	$2\frac{1}{2}$	$3\frac{1}{2}$	M / Term
$^4D_{\frac{1}{2}}$				-2	-1				$^4D_{\frac{1}{2}}$
$^4D_{1\frac{1}{2}}$			-3	-1	0	1			$^4D_{1\frac{1}{2}}$
$^4D_{2\frac{1}{2}}$		-4	-2	0	1	2	3		$^4D_{2\frac{1}{2}}$
$^4D_{3\frac{1}{2}}$	-5	-3	-1	1	2	3	4	5	
$\Sigma(M_L + 2M_S)$	-5	-7	-6	-2	2	6	7	5	$\Sigma(M_L + 2M_S)$

Fig. 7·18. The magnetic displacements of an erect ^{4}D term in weak and strong fields, Mg above and $(M_L + 2M_S)$ below. For a given value of M the sum is the same in both fields; this can also be verified in the previous figure.

The interpretation of this rule in terms of the model is somewhat obscure, but it clearly makes possible the calculation of g without the use of Landé's formula, that is, without the use of

* Pauli, *ZP*, 1923, **16** 155.

the quantum mechanics. Thus if the values of g for the $^4D_{\frac{1}{2}}$ to $^4D_{3\frac{1}{2}}$ components are written $g_{\frac{1}{2}}$ to $g_{3\frac{1}{2}}$, four equations are obtained for the four unknowns:

$$\tfrac{1}{2}(g_{\frac{1}{2}}+g_{1\frac{1}{2}}+g_{2\frac{1}{2}}+g_{3\frac{1}{2}})=2,$$
$$1\tfrac{1}{2}(g_{1\frac{1}{2}}+g_{2\frac{1}{2}}+g_{3\frac{1}{2}})=6,$$
$$2\tfrac{1}{2}(g_{2\frac{1}{2}}+g_{3\frac{1}{2}})=7,$$
$$3\tfrac{1}{2}(g_{3\frac{1}{2}})=5.$$

The last of these equations gives at once $g_{3\frac{1}{2}}=\frac{10}{7}$, in agreement with Landé's formula; and the other splitting factors may be deduced from the other equations.

If two quantities are identical in weak and strong fields, there is a strong presumption that they will be equal also in intermediate fields. And this is in agreement not only with such measurements as have been made, but also with Darwin's deductions from the quantum mechanics.

7. Invariance of the Γ sum

The displacements Γ, due to the coupling of the orbital and spin vectors, obey rules which are very similar to those which govern the magnetic displacements.

As Landé* first pointed out, if $\Gamma_{w.}$ be summed over all values of J and $\Gamma_{st.}$ over all values of M_S, while M, L and S are kept constant, then the two sums are equal:

$$\sum_J \Gamma_{w.}=\sum_{M_S}\Gamma_{st.};$$

though the displacement is determined in a weak field by the equation

$$\Gamma_{w.}=A\,.\,\tfrac{1}{2}\{J(J+1)-L(L+1)-S(S+1)\} \qquad \text{......(7·6)}$$

and in a strong field by

$$\Gamma_{st.}=A\,.\,M_L M_S. \qquad \text{......(7·7)}$$

Consider, for example, the ^4D term; in a weak field the displacements for $J=\frac{1}{2}$, $1\frac{1}{2}$, $2\frac{1}{2}$ and $3\frac{1}{2}$ are $-4\frac{1}{2}A$, $-3A$, $-\frac{1}{2}A$ and A respectively. Arranged by their permitted M values, these give the upper half of Fig. 7·19; in a strong field the displacements

* Landé, *ZP*, 1923, **19** 112.

may be obtained from Fig. 7·13, and inserted by their M values in the lower half of Fig. 7·19; the sums written at the foot of the two tables are identical.

Term \ M	$-3\frac{1}{2}$	$-2\frac{1}{2}$	$-1\frac{1}{2}$	$-\frac{1}{2}$	$\frac{1}{2}$	$1\frac{1}{2}$	$2\frac{1}{2}$	$3\frac{1}{2}$	M / Term
$^4D_{\frac{1}{2}}$				$-4\frac{1}{2}$	$-4\frac{1}{2}$				$^4D_{\frac{1}{2}}$
$^4D_{1\frac{1}{2}}$			-3	-3	-3	-3			$^4D_{1\frac{1}{2}}$
$^4D_{2\frac{1}{2}}$		$-\frac{1}{2}$	$-\frac{1}{2}$	$-\frac{1}{2}$	$-\frac{1}{2}$	$-\frac{1}{2}$	$-\frac{1}{2}$		$^4D_{2\frac{1}{2}}$
$^4D_{3\frac{1}{2}}$	3	3	3	3	3	3	3	3	$^4D_{3\frac{1}{2}}$
$\Sigma \frac{\Gamma w.}{A}$	3	$2\frac{1}{2}$	$-\frac{1}{2}$	-5	-5	$-\frac{1}{2}$	$2\frac{1}{2}$	3	$\Sigma \frac{\Gamma w.}{A}$

Term \ M	$-3\frac{1}{2}$	$-2\frac{1}{2}$	$-1\frac{1}{2}$	$-\frac{1}{2}$	$\frac{1}{2}$	$1\frac{1}{2}$	$2\frac{1}{2}$	$3\frac{1}{2}$	M / Term
$^4D_{\frac{1}{2}}$				$-1\frac{1}{2}$	-3				$^4D_{\frac{1}{2}}$
$^4D_{1\frac{1}{2}}$			0	0	$-\frac{1}{2}$	-1			$^4D_{1\frac{1}{2}}$
$^4D_{2\frac{1}{2}}$		$1\frac{1}{2}$	$\frac{1}{2}$	$-\frac{1}{2}$	0	$\frac{1}{2}$	1		$^4D_{2\frac{1}{2}}$
$^4D_{3\frac{1}{2}}$	3	1	-1	-3	$-1\frac{1}{2}$	0	$1\frac{1}{2}$	3	$^4D_{3\frac{1}{2}}$
$\Sigma \frac{\Gamma st.}{A}$	3	$2\frac{1}{2}$	$-\frac{1}{2}$	-5	-5	$-\frac{1}{2}$	$2\frac{1}{2}$	3	$\Sigma \frac{\Gamma st.}{A}$

Fig. 7·19. Displacements of an erect ^4D term in weak and strong fields, $\frac{1}{2}\{J(J+1)-L(L+1)-S(S+1)\}$ above and $M_L M_S$ below. For a given value of M the sum is the same in both fields.

The Γ sum rule makes possible the calculation of the Γ values of individual terms by processes of simple arithmetic, though without it these values can be obtained only by the methods of the quantum mechanics.

BIBLIOGRAPHY

The most thorough account is by Back in *Handbuch der Experimental Physik*, 1929, **22** 124.

CHAPTER VIII

ATOMIC MAGNETISM

1. Electronic theory

If a bar, hung in a uniform magnetic field, sets its length parallel to the field, it is said to be paramagnetic; while if it sets perpendicular to the field, it is diamagnetic. In 1895 Curie examined a large number of substances and showed that whereas the susceptibility of diamagnetic bodies is independent of the temperature, the susceptibility of paramagnetic bodies varies inversely as the temperature. For this striking difference Langevin* was able to account, when ten years later he developed the theory on which all more recent speculation has been founded.

This theory supposes that all substances are diamagnetic, though the diamagnetic properties may be masked by a superimposed paramagnetism; for all substances contain revolving electrons, and a magnetic field will make them precess about the axis of the field. This precession, already used to explain the Zeeman effect, implies a universal diamagnetism, for a precessing electric charge must behave as a magnet, and Lenz's law insists that this induced field shall be directed to oppose the applied field. This action is independent of the orientation of the atom, for it arises in the individual electronic orbits, so that the susceptibility should be independent of the temperature, as experiment shows it is.

Though this diamagnetism pervades all matter it can be observed only when the molecules are not themselves magnetic, for if they have a magnetic moment they will tend to set themselves parallel to the field and the resulting paramagnetism will mask it. The tendency of individual magnets to set themselves parallel to the field will be opposed by the random collisions of molecule with molecule; and as collisions become more violent the higher the temperature, the paramagnetic susceptibility should decrease as the temperature rises.

* Langevin, *J. de Phys.* 1905, **4** 678; *Ann. de Chim. et de P.* 1905, **5** 70.

The above argument is perfectly general and applies to all substances, but just for this reason it does not determine the size of the magnetic element. Other evidence, however, shows that this is molecular or larger, for magnetism is not in general an atomic property. Thus the susceptibility of a metal commonly changes abruptly on fusion, and even in the solid abrupt changes occur at various temperatures, corresponding presumably to modifications of the solid structure; tin is paramagnetic at one temperature, and diamagnetic at another. Again, the susceptibility of a compound cannot be adumbrated from the susceptibilities of its component atoms; iron carbonyl is diamagnetic, and so too is aluminium oxide, though derived from paramagnetic oxygen and paramagnetic aluminium; the Heusler bronzes are nearly as strongly magnetic as iron itself, though composed of aluminium, copper and manganese, all elements of small susceptibility.

Nevertheless, the magnetic properties of an atom can be investigated when it moves freely, as it does in a vapour at low pressure or as an ion in solution. Of the gases, only the inert gases and a few metallic vapours are monatomic, and so only these give atomic susceptibilities.

Postponing to the succeeding sections the experimental methods and results, consider how the magnetic properties of an atom are related to its electronic structure and to its spectrum.

In the single atom each electron orbit has a certain magnetic moment, and if the total angular momentum due to these orbits is J in units $h/2\pi$, then by the argument of equation (6·3) the magnetic moment should be

$$\frac{e}{2m_e c} \cdot \frac{h}{2\pi} \cdot J. \qquad \text{......(8·1)}$$

The Zeeman and Paschen-Back effects have taught, however, that the magnetic moment of an atom is not due only to electrons revolving about the nucleus, but in part to the inherent magnetism of the electron, so that the actual magnetic moment must be written, not J, but gJ in units $\frac{eh}{4\pi m_e c}$. This unit is commonly called the Bohr magneton; numerically it works out at

0·9153 erg gauss^{-1}; or if a mol is substituted for a single molecule, the magnetic moment becomes 5550 erg gauss^{-1} mol^{-1}.

The most obvious consequence of this argument is that whenever J is zero, the atom will be non-magnetic; and, in fact, the inert gases and all ions isoelectronic with them are diamagnetic. Further, since the Zeeman effect has shown that the orientation of J in the magnetic field is restricted by quantum conditions, the magnetic moment should be similarly restricted and should assume only the discrete values gM, where as usual

$$J \geqslant M \geqslant -J.$$

Beautiful confirmation of this theory is found in the work, initiated by Gerlach and Stern, on atomic rays. Measurements on ions in solution are also in agreement with theory; but they are considerably more difficult to interpret, since one measures only the average moment of ions which are being jostled.

2. Atomic rays*

In the experiments of Gerlach and Stern a beam of atoms, emitted from a small furnace, was first defined by two parallel slits and then passed between the poles of a powerful electromagnet, the pole pieces being arranged to give a very inhomogeneous field. Thus the lower pole piece is a bar cut in the shape of an inverted V and the upper is a parallel bar with a groove in it; the beam of atoms passes parallel to the two bars, between the lower edge and the upper groove (Fig. 8·1).

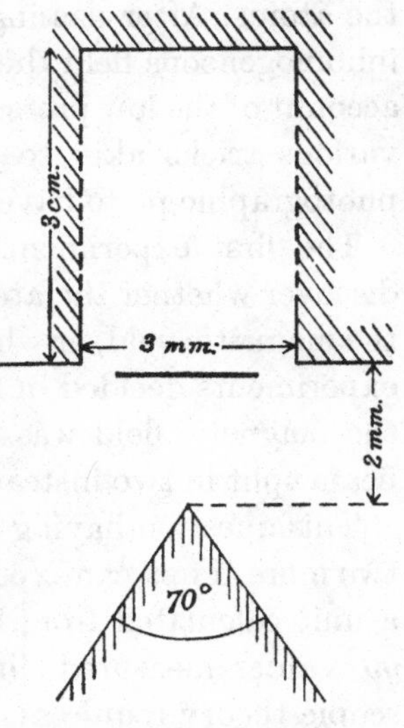

Fig. 8·1. The pole pieces used to deflect an atomic ray. The Hamburg set-up. After Fraser, *Molecular rays*.

This set-up is still in common use, but recently Rabi† has

* For a thorough account of this subject and especially of the experimental technique, Fraser, *Molecular rays*, 1931.

† Rabi, *ZP*, 1929, **54** 190.

developed another arrangement which is better adapted to absolute measurements of the Bohr magneton. In this set-up the atoms pass between the flat poles of a magnet, and as the deflection is found to depend only on the total change in the field strength, the latter need no longer be known from point to point.

The temperature of the furnace must be kept low so that the number of collisions between atoms after leaving the furnace may remain negligible; in some experiments on silver it was 1020° C. The temperature must also be accurately known, for it is used to determine the velocity of the atoms. After passing through the inhomogeneous field the atoms are received on a glass plate; on account of the low pressure the trace will often be very faint, but various tricks akin to the intensification of an underexposed photographic plate have been used to make the traces visible.

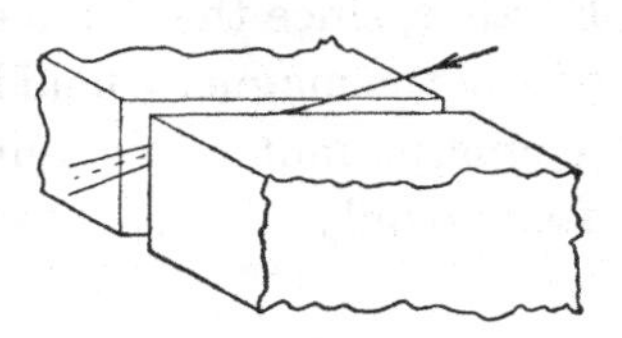

Fig. 8·2. Flat pole pieces used to deflect an atomic ray. The Rabi set-up. After Fraser, *Molecular rays*.

The first experiments, made about 1923, were directed to discover whether the atom sets itself in one of a few positions in the magnetic field, or whether all orientations are possible. These experiments decided in favour of the quantum theory, for when the magnetic field was applied to a beam of silver atoms, the beam split in two instead of merely growing wider.

This question having been decided, recent research has asked two more. First, can experiment substantiate the Bohr magneton, a unit calculated from independent data? And second, do the gJ values measured directly agree with those which spectroscopic theory requires?

The most accurate measurements of the Bohr magneton are those of Meissner and Scheffers,* who worked on lithium and potassium; both were shown to have a magnetic moment of 5552 erg. gauss^{-1} per gm. molecule, the probable error being $\pm \frac{1}{2}$ per cent. The theoretical value is 5550. Frase rconsiders that results accurate to 1 in 500 should be possible.†

* Meissner and Scheffers, *PZ*, 1933, **34** 48.

† Fraser, *Molecular rays*, 1931, 138.

The second question can best be answered by reviewing Mendeléeff's classification column by column, but first the precise scope of the method must be clarified. It has often been supposed that the Gerlach and Stern technique will determine the product gJ, but in fact it can measure this product only if J is known. For consider an atom in a state having $J = 1\frac{1}{2}$; if all atoms emitted by the furnace have the same velocity, or in a convenient phrase, if the rays are monochromatic, then the pattern would consist of four equally spaced traces having $M = \pm\frac{1}{2}, \pm 1\frac{1}{2}$.* But as the rays are not monochromatic, each trace is a band not a sharp line, and the two traces on one side of the centre sum to a rather broad single trace as Fig. 8·3 shows.

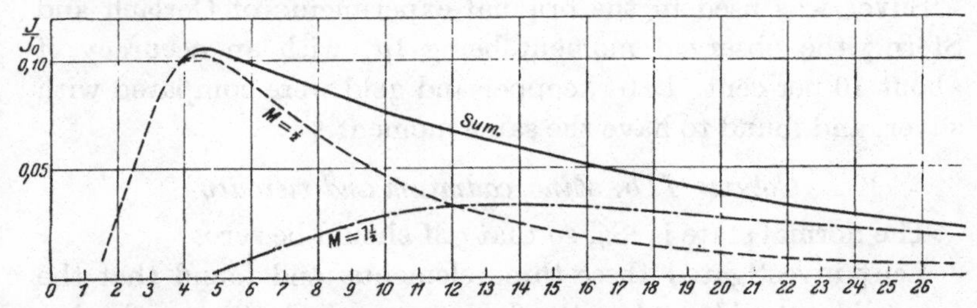

Fig. 8·3. Theoretical traces for atoms having $M = \frac{1}{2}$ ——— and $M = 1\frac{1}{2}$ —·—·—; the result of summing these, shown full, is the theoretical trace for an atom having $J = 1\frac{1}{2}$. The vertical intensity scale and the horizontal deflection scale are in arbitrary units. After Fraser, *Molecular rays*.

Now the summation trace, shown by a continuous line in Fig. 8·3, has only one maximum, and so cannot be distinguished from a trace due to atoms having $J = \frac{1}{2}$ and a different value of g. But if J is known and if the beam does not contain any atoms in metastable states, the magnetic deflection determines g uniquely.

If, however, metastable states are present in the beam, the J values of these states are not sufficient to allow the summation trace to be interpreted. Besides the J values, one must know in what proportions the various states occur in the beam; fortunately gas theory allows these to be calculated from the temperature of the source. This method will be illustrated in the discussion of oxygen.

* Rabi and Cohen, *PR*, 1933, **43** 582*a*, have used a monochromatic beam to measure the nuclear spin of sodium.

Hydrogen

The normal state is $^2S_{\frac{1}{2}}$, so that a gM value of ± 1 is indicated; measurements by Wrede* confirm this with an estimated error of 4 to 5 per cent.

Column Ia. The alkali metals

Sodium and potassium have been examined by Leu,† who found a moment of $1\mu_B$ for both. Taylor‡ compared lithium with potassium, and showed that both have the same magnetic moment.

Column Ib. Copper, silver and gold

Again the normal state is $^2S_{\frac{1}{2}}$, giving a gM value of ± 1.

Silver was used in the original experiments of Gerlach and Stern,§ the observed moment being $1\mu_B$ with an accuracy of about 10 per cent. Later, copper and gold were compared with silver, and found to have the same moment.||

Column IIb. Zinc, cadmium and mercury

The normal state is $^1S_0$, so that gM should be zero.

Leu† investigated these three elements, and found that the trace did not widen when the field was applied, thus confirming the prediction of theory.

Column III. Thallium

The normal state of thallium is an erect ^2P term with an interval of 7792 cm.$^{-1}$

Gas theory asserts that if atoms exist in two states having energies E_1, E_2 and momenta J_1, J_2, the numbers present in these states are in the ratio of

$$(2J_1+1)\,e^{-E_1/kT} : (2J_2+1)\,e^{-E_2/kT}, \qquad \text{......(8·2)}$$

where k is Boltzmann's constant; that is, the $^2P_{\frac{1}{2}}$ and $^2P_{1\frac{1}{2}}$ states should exist in the ratio of $1 : 2e^{-ch\Delta\nu/kT}$, where $\Delta\nu$ is the doublet interval, measured in wave-numbers.

In Leu's experiments the temperature was 980 A., and at this temperature the above ratio works out as $1 : 3{\cdot}5 \,.\, 10^{-5}$; so that a

* Wrede, *ZP*, 1927, **41** 569.
† Leu, *ZP*, 1927, **41** 551.
‡ Taylor, J. B., *ZP*, 1929, **52** 846.
§ Gerlach and Stern, *AP*, 1924, **74** 673.
|| Gerlach, *AP*, 1925, **76** 163.

PLATE IV. MAGNETIC DEFLECTION OF ATOMIC RAYS

1. Traces obtained by Gerlach and Stern in 1921 using a beam of silver atoms and a circular aperture. (*a*) shows the trace without field; (*b*) the beam split in two by the field.

2. Traces of a lithium beam split in two by a magnetic field. Comparison with Fig. 1 shows the improved technique; slits have been substituted for circular apertures.

3. Deflection pattern of potassium obtained by Rabi using flat pole pieces. (*ZP*, 1929, **54** 190.)

4. Deflection pattern of oxygen obtained by Kurt and Phipps; an undeflected trace and two deflected traces appear. (*PR*, 1929, **34** 1357.)

5. Deflection traces of bismuth at slit temperatures of 910 and 1155° C. The undeflected trace visible at the lower temperature is ascribed to undissociated bismuth molecules; at the higher temperature the pattern consists of two components, not four although the ground state is $^4S_{1\frac{1}{2}}$.

These photographs have been borrowed from *Molecular Rays*, 1931, by R. G. J. Fraser.

Plate IV

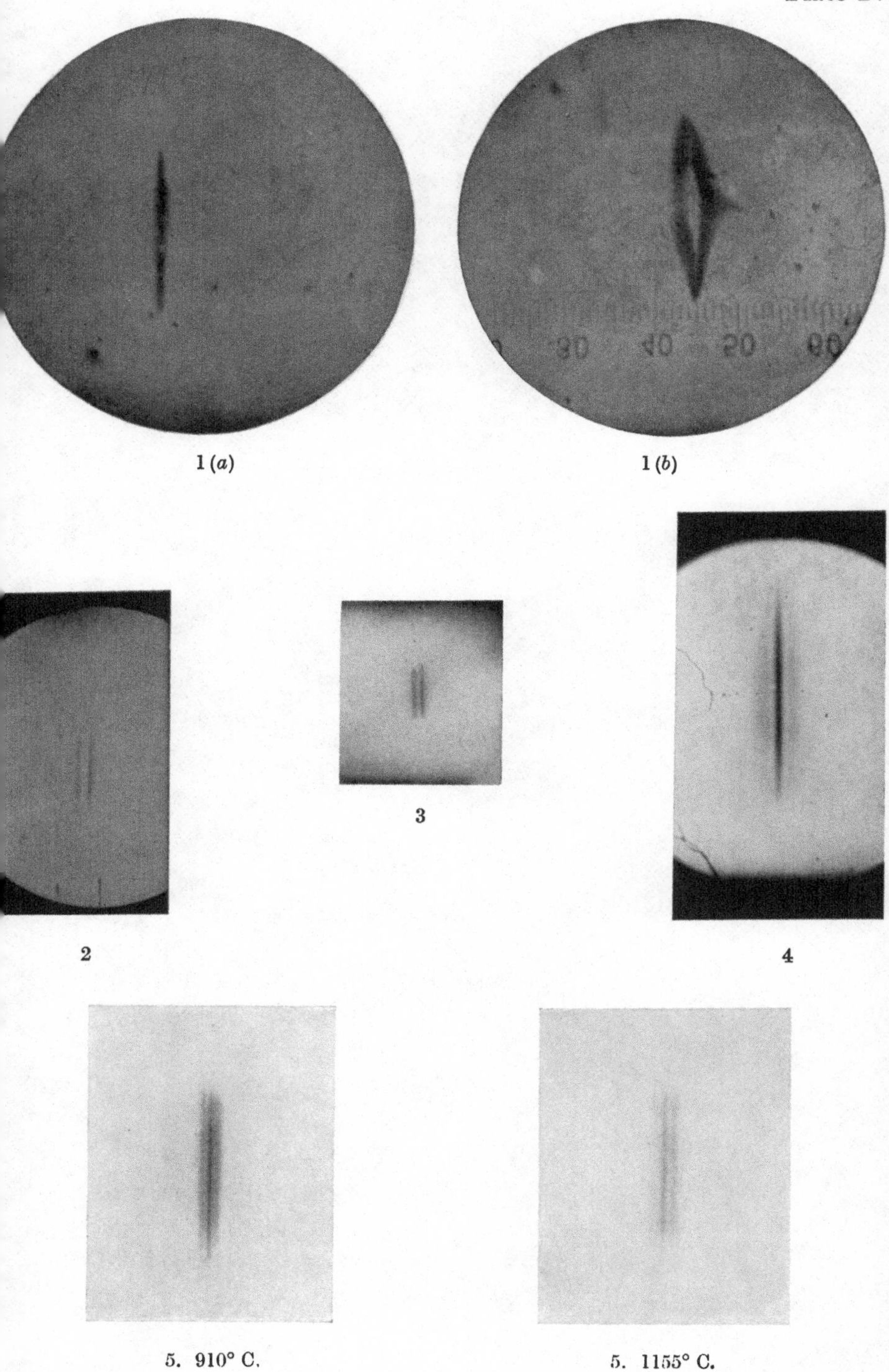

1 (*a*) 1 (*b*)

2 3 4

5. 910° C. 5. 1155° C.

negligible number of atoms are in the $^2P_{1\frac{1}{2}}$ state, and the splitting should be that predicted for the $^2P_{\frac{1}{2}}$ level alone, namely $Mg = \pm \frac{1}{3}$; and this is exactly what Leu* found.

Column IV. Tin and lead

The normal state of tin and lead is $^3P_0$; moreover, the intervals are large, 1692 cm.$^{-1}$ in tin and 7817 cm.$^{-1}$ in lead, so that only a small proportion of the atoms will be in the metastable $^3P_1$ state.

Gerlach† investigated the two metals and found that both have zero magnetic moment in agreement with this prediction.

Column V. Bismuth

The normal state of bismuth is $p^3\,{}^4S_{1\frac{1}{2}}$,‡ so that the beam should consist of atoms having four moments, namely $\pm \frac{1}{2}g$ and $\pm 1\frac{1}{2}g$; but, as explained above, theory suggests that these will produce a symmetrical pattern consisting of only two components. In an early research Gerlach obtained an undeflected trace, but this seems to have been due to diatomic molecules; Leu and Fraser,§ using a strongly superheated beam, obtained just the two components predicted.

Evaluation of the deflected traces gives g a value of 1·45. Now a $p^3\,{}^4S_{1\frac{1}{2}}$ term should have $g = 2$ if the coupling is Russell-Saunders, but $g = \frac{4}{3}$ if the coupling is of the (jj) type,|| and as bismuth is a heavy element the coupling may be expected to approximate to the latter. Indeed, the $p^3\,{}^2D_{1\frac{1}{2}}$ state has been found to have a g value of 1·224¶ instead of the 0·8 or 1·47 which the Russell-Saunders and (jj) couplings respectively require. Thus experiment is in satisfactory agreement with theory, but further work on the Zeeman effect of bismuth is much to be desired.

This investigation of bismuth illustrates at once the power and limitations of the method of atomic rays; since only two traces were observed, the experiment cannot be used to determine

* Leu, *ZP*, 1927, **41** 551.
† Gerlach, *AP*, 1925, **76** 163.
‡ Cf. chapter XIII on the combination of several electrons.
§ Leu and Fraser, *ZP*, 1928, **49** 498.
|| Cf. chapter XVIII on abnormal coupling.
¶ Back and Goudsmit, *ZP*, 1928, **47** 174.

the J value of the ground term; that must be obtained from work on spectra; but given the moment of the ground state the method is able to determine the magnetic moment.

Column VI. Oxygen

The normal state of the oxygen atom is an inverted ^3P level, with energy values of

$^3P_2$	$^3P_1$	$^3P_0$
0	158·1	226·8 cm.$^{-1}$

With such small intervals all three states are to be expected in the beam, calculation by the method used for thallium showing that at a temperature of 613 A. the relative probabilities are

$$^3P_2 : {}^3P_1 : {}^3P_0 :: 5 : 2{\cdot}07 : 0{\cdot}59.$$

The possible Mg values for each of the three states are ± 3, $\pm 1\frac{1}{2}$, 0; $\pm 1\frac{1}{2}$, 0; and 0. The three Mg values of 3, $1\frac{1}{2}$ and 0 should therefore be present in the ratio

$$(3) : (1\tfrac{1}{2}) : (0) :: 0{\cdot}438 : 0{\cdot}741 : 1.$$

Fig. 8·4 shows the theoretical intensity curves for Mg values of $1\frac{1}{2}$ and 3 dotted in and labelled B and C; they are drawn with reference to the observed hydrogen curve, A, as standard. The summation curve D has but a single maximum, and this corresponds to an average magnetic moment of $1{\cdot}71\mu_B$. Experiment is in admirable agreement;* the deflection pattern of oxygen consists of an undeflected trace, and two nearly symmetrical deflected traces. The deflected traces were measured against similar traces for atomic hydrogen, the latter being assumed to have a moment of $1\mu_B$; the moment indicated by oxygen traces was then $1{\cdot}67\mu_B$.

Column VIII. Iron, cobalt and nickel

Iron, cobalt and nickel have been investigated by Gerlach and his co-workers. At first iron vapour was supposed to be non-magnetic, but more recently Gerlach† has interpreted the observed pattern as showing atoms of many different magnetic

* Kurt and Phipps, *PR*, 1929, **34** 1357.

† Gerlach, *J. de Phys.* 1929, **10** 274.

moments. And the same seems to be true of cobalt, save that the evidence limits the maximum moment to $6\mu_B$.

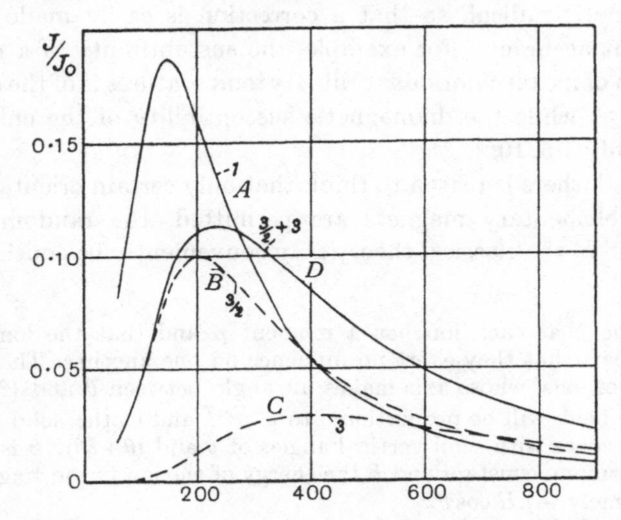

Fig. 8·4. Deflection pattern of oxygen. *A* is an empirical curve for hydrogen, in which $Mg=1$; *B* and *C* are theoretical curves for atoms having *Mg* values of $1\frac{1}{2}$ and 3 drawn against the hydrogen curve as standard. *D*, the sum of *B* and *C*, is thus the theoretical trace for oxygen. The intensity and deflection scales are in arbitrary units. After Fraser, *Molecular rays*.

But the traces obtained with cobalt, like those obtained with nickel, have not yet been satisfactorily explained; for the patterns apparently show resolved traces due to atoms of different moments; and this is contrary alike to the thorough analysis given by Fraser* and to the experimental work on bismuth and oxygen.

3. Paramagnetic ions

Solutions of certain salts, such as nickel sulphate and chloride, have susceptibilities which are proportional to the concentration, so that the susceptibility per mol is easily calculated. In other solutions whose susceptibility is not proportional to the concentration, the susceptibility of the ions may be obtained by

* Fraser, *Molecular rays*, 1931, 132.

extrapolating to infinite dilution. Moreover, the susceptibility of a paramagnetic ion is always large compared with that of a diamagnetic radical, so that a correction is easily made with sufficient accuracy. For example, the susceptibility of a molal solution of nickel chloride at ordinary temperatures is of the order of 5.10^{-3}, while the diamagnetic susceptibility of the chlorine ion is only $1{\cdot}5.10^{-5}$.

Though there is reason to think that only certain orientations of the elementary magnets are permitted, the random distribution of the classical theory may conveniently be considered first.*

Assume that each ion has a moment μ and that the ions are so far apart that they exert no influence on one another. Then the number of ions, whose axis makes an angle between θ and $(\theta+\delta\theta)$ with the field, will be proportional to $e^{-E/kT}$ and to the solid angle between cones with semi-vertical angles of θ and $(\theta+\delta\theta)$. k is here the Boltzmann constant and E the energy of the ion in the magnetic field, namely $-\mu H\cos\theta$.

If the volume under consideration contains n molecules,

$$n=\int_0^\pi Ce^{\frac{\mu H\cos\theta}{kT}}\,2\pi\sin\theta\,d\theta$$

$$=2\pi C\int_0^\pi e^{a\cos\theta}\sin\theta\,d\theta, \qquad \text{......(8·3)}$$

where $a=\dfrac{\mu H}{kT}$. And the mean effective magnetic moment $\bar{\mu}$ of these ions is given by the equation

$$n\bar{\mu}=2\pi C\int_0^\pi \mu\cos\theta\,.\,e^{a\cos\theta}\sin\theta\,d\theta. \qquad \text{......(8·4)}$$

Dividing the second of these equations by the first,

$$\bar{\mu}=\mu\frac{\int_0^\pi e^{a\cos\theta}\cos\theta\sin\theta\,d\theta}{\int_0^\pi e^{a\cos\theta}\sin\theta\,d\theta}. \qquad \text{......(8·5)}$$

At ordinary temperatures a is always very small, for even if an atom has a moment of $5\mu_B$ in a field of 4.10^4 gauss, then a is still only $4{\cdot}7.10^{-2}$.

* Stoner, *Magnetism and atomic structure*, 1926, 71.

Accordingly, write $e^{a\cos\theta}=1+a\cos\theta$.

Then
$$\bar{\mu}=\mu\frac{\int\cos\theta\, d(\cos\theta)+a\int\cos^2\theta\, d(\cos\theta)}{\int d(\cos\theta)+a\int\cos\theta\, d(\cos\theta)},$$

but
$$\int_0^\pi \cos\theta\, d(\cos\theta)=0,$$

so that
$$\bar{\mu}=\mu a\frac{\int_0^\pi \cos^2\theta\, d(\cos\theta)}{\int_0^\pi d(\cos\theta)}$$
$$=\mu a\,\overline{\cos^2\theta}, \qquad \ldots\ldots(8{\cdot}6)$$

where $\overline{\cos^2\theta}$ denotes the mean value of $\cos^2\theta$ over the surface of a sphere.

Thus
$$\bar{\mu}=\mu\frac{\mu H}{3kT}. \qquad \ldots\ldots(8{\cdot}7)$$

The magnetic moment induced per mol is however the molal susceptibility χ, so that
$$\chi=\frac{N\bar{\mu}}{H}=\frac{N^2\mu^2}{3RT}, \qquad \ldots\ldots(8{\cdot}8)$$

where N is the number of atoms per mol and $R=Nk$ is the gas constant. Thus theory accounts for the Curie law, which states that paramagnetic susceptibility varies inversely as the absolute temperature.

Accordingly, if the atom may take up any orientation under the dual influence of the magnetic field and collisions with other atoms, its magnetic moment is
$$\mu=\frac{1}{N}\sqrt{3RT\chi}. \qquad \ldots\ldots(8{\cdot}9)$$

This completes the classical theory. The vector model allows, not any orientation, but only those determined by the quantum number M. Equation 8·6 however still holds, and as the mean value of $\cos^2\theta$ is simply the mean value of M^2/J^2,
$$\mu=\mu\frac{\mu H}{kT}\cdot\frac{\overline{M^2}}{J^2}. \qquad \ldots\ldots(8{\cdot}10)$$

In the vector model,
$$\frac{\overline{M^2}}{J^2}=\frac{1}{J^2(2J+1)}\sum_{-J}^{J}M^2=\frac{J(J+1)}{3J^2}, \qquad \ldots\ldots(8{\cdot}11)$$

so that
$$\chi=\frac{N\bar{\mu}}{H}=\frac{J+1}{3J}\cdot\frac{N^2\mu^2}{RT}. \qquad \ldots\ldots(8{\cdot}12)$$

But spectral theory shows that the magnetic moment of the atom is $gJ\mu_B$, where μ_B is the Bohr magneton. Thus

$$\chi=\frac{g^2 J(J+1)}{3}\cdot\frac{N^2\mu_B{}^2}{RT} \qquad \text{......(8·13)}$$

or

$$g\sqrt{J(J+1)}\,\mu_B=\frac{1}{N}\sqrt{3RT\chi}. \qquad \text{......(8·14)}$$

The above interpretation in terms of the vector model is open to more than one criticism, but the result is confirmed by the wave mechanics.*

Theory thus shows that if the paramagnetic susceptibility χ of a solution is due to ions of magnetic moment μ, which may take up any orientation, then the magnitude of the moment is

$$\mu=\frac{1}{N}\sqrt{3RT\chi}. \qquad \text{......(8·9)}$$

Or if the ions have moments $gJ\mu_B$, and may only orientate themselves in one of the positions specified by the usual quantum conditions, then

$$g\sqrt{J(J+1)}\,\mu_B=\frac{1}{N}\sqrt{3RT\chi}. \qquad \text{......(8·14)}$$

This theory is tested by applying it to three groups of elements, monatomic vapours, solutions of the rare earths and solutions of the iron row. It should be applicable also to the palladium and platinum rows, but there is very little experimental data.

The only monatomic gases on which measurements seem to have been made are the inert gases and potassium vapour. The inert gases are all diamagnetic, as might be expected since their ground state is ${}^1S_0$, but potassium vapour† is paramagnetic, the value of $\frac{1}{N}\sqrt{3RT\chi}$ being $\sqrt{3}\mu_B$, as it should be if the ground state is ${}^2S_{\frac{1}{2}}$.

The rare earths all form in solution ter- or tetravalent ions; their ground states are not known, but Hund has supposed that all electrons outside the xenon core occupy f orbits, and if they do, the ground terms can be calculated; for Hund's energy rules state that of all terms produced by the f^n configuration the ground

* Van Vleck, *PR*, 1927, **29** 727; 1928, **31** 587.

† Gerlach, *J. de Phys.* 1929, **10** 273.

term has the greatest spin and orbital vectors; when the shell is more than half full the terms are inverted.*

This theory, as Figs. 8·5 and 8·6 show, agrees very well with the experimental values obtained for all but two of the salts, but the values predicted for samarium and europium are outside the

Ion	No. of f electrons	Ground term	g	$g\sqrt{J(J+1)}$	Experimental moments in Bohr magnetons		
					C.	St. M.	de H.
La^{3+}	0	$^1S_0$	$\frac{0}{0}$	0	—	Diamagnetic	—
Ce^{4+}	0	$^1S_0$	$\frac{0}{0}$	0	—	0·16	—
Ce^{3+}	1	$^2F_{2\frac{1}{2}}$	$\frac{6}{7}$	2·54	2·32	—	—
Pr^{4+}	1	$^2F_{2\frac{1}{2}}$	$\frac{6}{7}$	2·54	—	2·79	—
Pr^{3+}	2	$^3H_4$	$\frac{4}{5}$	3·57	3·60	3·47	—
Nd^{3+}	3	$^4I_{4\frac{1}{2}}$	$\frac{8}{11}$	3·61	3·62	3·51	—
Il^{3+}	4	$^5I_4$	$\frac{3}{5}$	2·68	—	—	—
Sm^{3+}	5	$^6H_{2\frac{1}{2}}$	$\frac{2}{7}$	0·85	1·54	1·42	—
Eu^{3+}	6	$^7F_0$	$\frac{0}{0}$	0	3·62	3·63	—
Gd^{3+}	7	$^8S_{3\frac{1}{2}}$	2	7·94	8·09	8·12	—
Tb^{3+}	8	$^7F_6$	$\frac{3}{2}$	9·73	9·52	9·05	—
Dy^{3+}	9	$^6H_{7\frac{1}{2}}$	$\frac{4}{3}$	10·63	10·51	10·67	—
Ho^{3+}	10	$^5I_8$	$\frac{5}{4}$	10·59	10·48	10·46	—
Er^{3+}	11	$^4I_{7\frac{1}{2}}$	$\frac{6}{5}$	9·58	9·46	9·41	9·00
Tu^{3+}	12	$^3H_6$	$\frac{7}{6}$	7·56	7·18	7·55	—
Yb^{3+}	13	$^2F_{3\frac{1}{2}}$	$\frac{8}{7}$	4·52	4·40	4·53	—
Lu^{3+}	14	$^1S_0$	$\frac{0}{0}$	0	Diamagnetic	—	—
Hf^{4+}	14	$^1S_0$	$\frac{0}{0}$	0	Diamagnetic	—	—

Fig. 8·5. Magnetic moments of the rare earths. The experimental values are from the work of Cabrera (*J. de Phys.* 1922, **3** 443) and Stefan Meyer (*PZ*, 1925, **26** 51), assuming 4·96 Weiss magnetons to 1 Bohr magneton. The third value of Er^{3+} is due to de Haas, Wiersma and Capel, *K. Akad. Amsterdam*, 1929, **32** 738.

possible error of experiment. Recently, however, Freed has shown that at low temperatures the susceptibility of samarium approaches the value predicted by Hund; the change which occurs when the temperature rises is due to the passage of some ions into states with larger moments; of these states there is spectroscopic evidence.† Moreover, Van Vleck‡ has derived a formula for paramagnetic susceptibility from the quantum mechanics, and

* This point is developed in Chapter XIII.

† Freed and Spedding, *N*, 1929, **123** 525.

‡ Van Vleck and Frank, *PR*, 1929, **34** 1495, 1625.

obtained values which agree better with experiment than Hund's simple theory, namely $1{\cdot}66\mu_B$ for Sm^{3+} and $3{\cdot}53\mu_B$ for Eu^{3+}.

The theory that the ground states of the rare earths contain only f electrons is thus confirmed, but a similar postulate applied

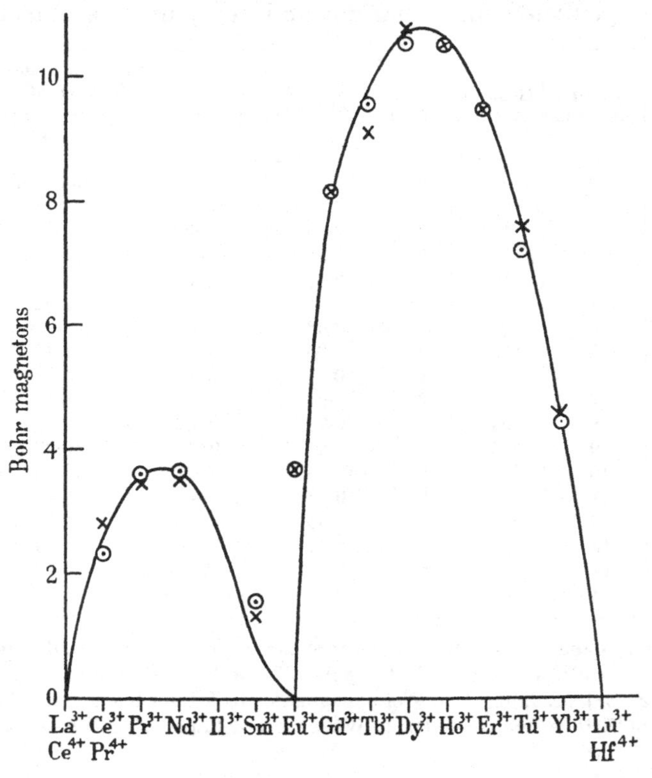

Fig. 8·6. Magnetic moments of the rare earths; the curve gives the predictions of theory, the points the experimental values of Cabrera ⊙ and S. Meyer ×. After Laporte and Sommerfeld, *ZP*, 1926, **40** 338.

to the iron row is much less satisfactory. However, one fact at least is clear, the susceptibilities are dependent only on the number of electrons and not at all on the atomic number; K^+, Ca^{2+}, Sc^{3+}, Ti^{4+} are all diamagnetic, while V^{2+}, Cr^{3+} and Mn^{4+} are all paramagnetic and yield for $\frac{1}{N}\sqrt{3RT\chi}$ values between 3·8 and $4{\cdot}0\mu_B$.

If the ions of the iron row contain only 3d electrons outside the argon shell, and Hund's energy rules are valid, the magnetic moments can be calculated. In fact prediction does not agree well with the empirical values, and various explanations have been sought. As the intervals are smaller than in the rare earths, some meta-stable states may be present; so instead of assuming the intervals infinite Laporte and Sommerfeld* calculated the moments on the second limiting assumption that the intervals of the ground terms are zero (Figs. 8·7–8·8).

Ions	No. of d electrons	Ground term	$g\sqrt{J(J+1)}$		Experimental moments in Bohr magnetons	$2\sqrt{S(S+1)}$
			$\Delta\nu=\infty$	$\Delta\nu=0$		
K^+, Cu^{2+}, Sc^{3+}, Ti^{4+}	0	$^1S_0$	0	0	Diamagnetic	0
V^{4+}	1	$^2D_{1\frac{1}{2}}$	1·55	2·75	1·7	1·73
V^{3+}	2	$^3F_2$	1·63	4·03	2·4	2·83
V^{2+}, Cr^{3+}, Mn^{4+}	3	$^4F_{1\frac{1}{2}}$	0·77	4·52	3·8, 4·0	3·87
Cr^{2+}, Mn^{3+}	4	$^5D_0$	0	4·85	4·8	4·90
Mn^{2+}, Fe^{3+}	5	$^6S_{2\frac{1}{2}}$	5·92	5·92	5·8–5·9	5·92
Fe^{2+}	6	$^5D_4$	6·71	4·85	5·2–5·3	4·90
Co^{2+}	7	$^4F_{4\frac{1}{2}}$	6·63	4·52	4·8–5·0	3·87
Ni^{2+}	8	$^3F_4$	5·59	4·03	3·2–3·4	2·83
Cu^{2+}	9	$^2D_{2\frac{1}{2}}$	3·55	2·75	1·8–2·0	1·72
Cu^+, Zn^{2+}	10	$^1S_0$	0	0	Diamagnetic	0

Fig. 8·7. Magnetic moments of the first twelve elements of the iron row.

The two curves thus obtained coincide at three points, namely at Cu^{2+}, Mn^{2+} and Zn^{2+}, where the ground states are $^1S_0$, $^6S_{2\frac{1}{2}}$ and $^1S_0$; and at these three points experiment confirms the values predicted. For the rest the experimental curve may be expected to lie between the two curves, one giving the atomic moments if the ground states have $\Delta\nu=\infty$ and the other the moments if $\Delta\nu=0$. In fact all but two of the experimental values do satisfy this condition, but these two—for the cupric and nickelous ions —are very difficult to explain away.

There is, however, one other consideration which has been too often forgotten. Ions in solution are usually hydrated; of this there is abundant evidence; and accordingly the moments which are measured are the moments of hydrated, and not of the free, ions. Thus the chromic ion in solution is the complex $Cr(H_2O)_6{}^{3+}$

* Laporte and Sommerfeld, *ZP*, 1926, **40** 333.

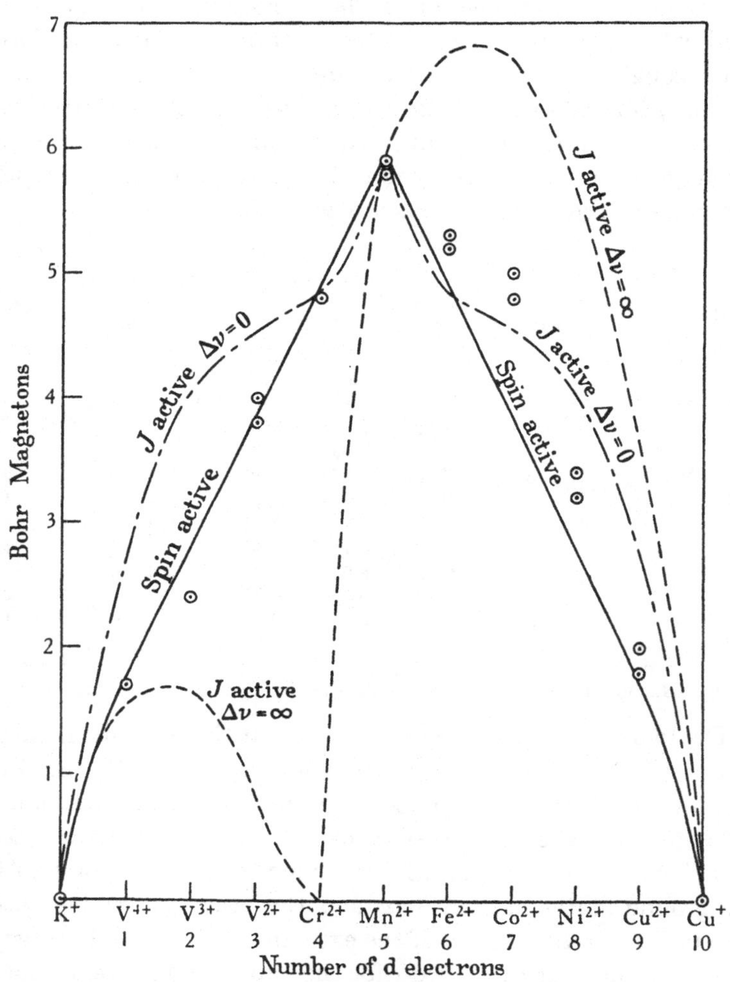

Fig. 8·8. Magnetic moments of the first twelve elements of the iron row. The points are empirical; the curves give the predictions of theory according as

(*a*) the spin alone is active ————;

(*b*) J is active and the splitting of the ground terms wide — — —;

(*c*) J is active and the splitting narrow —·—·—.

and not the atomic Cr^{3+} occurring in a vapour.* If the two are identical in the rare earth elements, it is only because in them the incomplete electronic shell lies deep within the atom, and so remains uninfluenced by any water molecules which attach themselves to the exterior. Van Vleck† suggested that this presence of an ionic atmosphere might restrict the freedom of the orbital vector, while leaving the spin vector free; if this is true then the paramagnetic susceptibility should be a function of S instead of J, and instead of equation (8·14), we should have

$$g(S)\sqrt{S(S+1)}\,\mu_B=\frac{1}{N}\sqrt{3RT\chi}, \qquad (8\cdot 15)$$

where $g(S)$ is the magnetic splitting factor of the spin, namely 2. Actually this equation gives better agreement than the other hypotheses, but the orbital vector still contributes something, and deviations from equation (8·15) are therefore observed.‡

What little work has been done on the palladium and platinum rows serves to emphasise this warning; the magnetic moments of the Ru^{3+} and Ir^{4+} ions, which should both have $^6S_{2\frac{1}{2}}$ ground terms, are $1{\cdot}98\mu_B$ and $2{\cdot}71\mu_B$ instead of the $5{\cdot}88\mu_B$ which theory dictates; and this is the more surprising as this particular value is accurately confirmed by measurements on Mn^{2+} and Fe^{3+}. But Guthrie and Bourland,§ in what seems to be the only recent paper on these rows, go on to show that some compounds do not even obey the Curie law, so that the simple theory reveals itself as wholly inadequate.

BIBLIOGRAPHY

Magnetism and atomic structure, 1926, by E. C. Stoner, is still valuable in spite of much recent work; a revision, *Magnetism and matter*, 1934, appeared after the present manuscript was complete. *Molecular rays*, 1931, by R. Fraser, is a keenly critical account of a subject previously much misunderstood; but Rabi and his co-workers have made considerable advances in technique since 1931; some reference to this is made in chapter XX on hyperfine structure, since Rabi's work has been recently directed to the measurement of nuclear moments.

* Freed, *Am. Chem. Soc. J.* 1929, **49** 2456.

† Van Vleck, *Theory of electric and magnetic susceptibilities*, 1932, 284 f., where a general discussion is given.

‡ Janes, *PR*, 1935, **48** 78.

§ Guthrie and Bourland, *PR*, 1931, **37** 303.

CHAPTER IX

STARK EFFECT

1. Experimental

The effect of an electric field on the emission of spectral lines was first demonstrated by Stark in 1913,* and in the following years he brilliantly measured the fine structure and polarisation in hydrogen and some other elements. Though the splitting of the hydrogen lines was explained by the quantum theory as early as 1916, the patterns produced in other elements have not led to

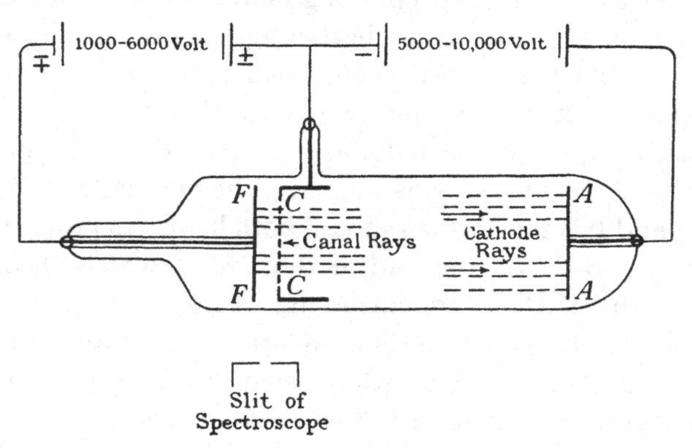

Fig. 9·1. Discharge tube used to observe the transverse Stark effect. After Andrade, *The structure of the atom.*

results of general interest, so that the Stark effect has remained a backwater, unvisited by most spectroscopists. In contrast the Zeeman patterns are in the main stream of progress.

That the electric field would influence the emission of spectral lines had been predicted in the early years of the century, when the Zeeman effect was being worked out, but experiment has always been extremely difficult; a field of 100,000 volts/cm. cannot be applied to a Geissler tube, because Geissler tubes are comparatively good conductors and the field simply collapses.

* Stark, *AP*, 1914, **43** 965, 983. For a recent account see Stark, *Hb. d. Expt. Phys.* 1927 **21**.

Stark therefore used the light emitted by a canal ray tube in the layer just behind the perforated cathode (Fig. 9·1). By placing a second electrode F parallel and close to the cathode, he was able to produce a uniform and measurable field in a space of a few millimetres; a very large potential gradient can be maintained, because the free path of the ions is much greater than the distance between the plates, so that very few ions are produced by collision.

Shortly after Stark's long efforts had met with success, Lo Surdo showed that the splitting is visible in front of the cathode, where the canal rays fall through the strong field, which exists between the negative glow and the cathode. The intensity is much greater in front than behind, so that the Lo Surdo technique is commonly used in analysis, but it is not so well adapted to accurate measurement.*

2. Splitting of hydrogen lines

In an electric field the lines of the Balmer series split into a number of components, and these are arranged symmetrically about the position of the undisplaced line. When viewed along a line normal to the electric field, all components appear polarised either parallel or perpendicular to the field just as in the Zeeman effect; and nothing new is learnt by viewing a new pattern along the axis of the field, for the π components are then invisible and the σ components are unpolarised.

The displacements of the components are proportional to the applied field† and are simple multiples of a fundamental interval; moreover, when expressed in wave-numbers this interval is the same for all lines, being 6·45 cm.$^{-1}$ in a field of 100,000 volt/cm. Provided with this key the structure of the four visible lines of hydrogen may be read from Fig. 9·2; this figure draws attention also to two other regularities; the number of components increases with the series number of the line, while the intense π components lie in general on the outside of the pattern, and the intense σ components towards the centre.

* Lo Surdo, *Rend. d. Lincei*, 1913, **22** 664; 1914, **23** 82, 143, 252, 326.

† Stark, *AP*, 1915, **48** 193. For a small deviation from proportionality in intense fields, see Traubenberg and Gebauer, *ZP*, 1919, **54** 307; **56** 254; Ishida and Hiyama, *Inst. Phys. and Chem. Tokyo Sci. P.* 1928, **9** 1.

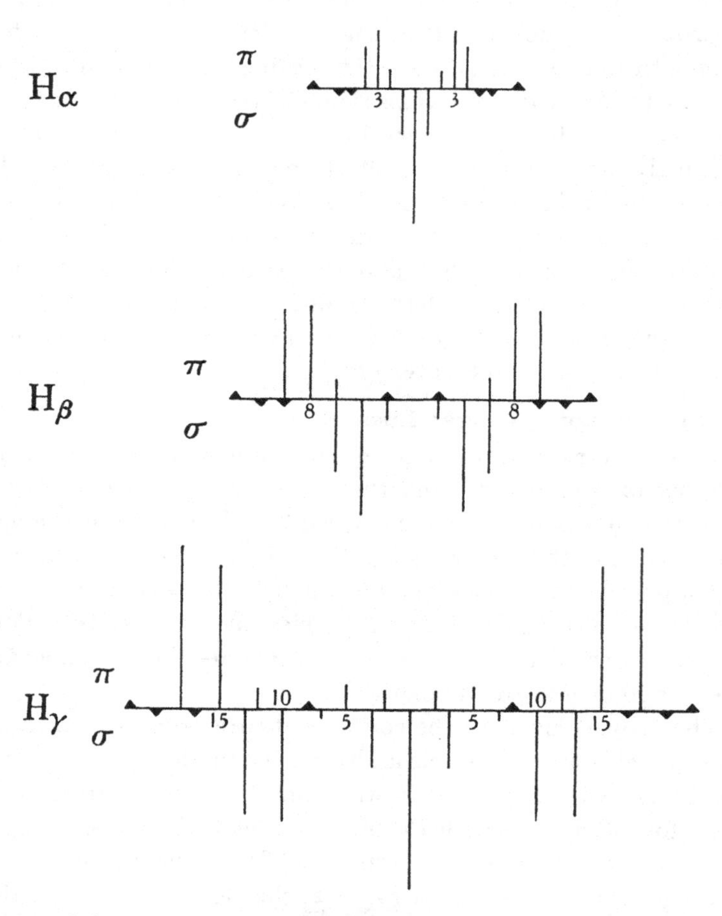

Fig. 9·2. Splitting of the Balmer lines of hydrogen in an electric field. These patterns are based on Schrödinger's theory, which Foster's observations confirm very closely. A triangle indicates a component so weak that if drawn to scale it would be invisible.

Theory accounts for these patterns very satisfactorily.* In an electric field, a hydrogen level is displaced by

$$\Delta E/ch = \frac{3hF}{8\pi^2 m_e eZc} ns,$$

where s is defined by the equation

$$s = n - 1 - 2l + m_l.$$

And m_l assumes a series of integral values, subject to the condition

$$0 \leqslant m_l \leqslant l.$$

In these equations m_e and e are the mass and charge of the electron; n is the chief quantum number, which determines the

n	l	m_l	s	ns
3	0	0	2	6
	1	0	0	0
		1	1	3
	2	0	−2	−6
		1	−1	−3
		2	0	0

Fig. 9·3. Quantum numbers and energies of the Stark levels of a hydrogen atom having $n=3$.

major axis, and F is the field strength in e.s.u. h, Z and c are Planck's constant, the atomic number and the velocity of light, respectively. Thus in a hydrogen-like atom all levels and all lines will be displaced by a simple multiple of the fundamental interval $\dfrac{3hF}{8\pi^2 m_e eZc}$, and it is just this interval which experiment reveals. A recent measurement gives the interval a numerical value of 6·44 cm.$^{-1}$ in a field of 100,000 volt/cm., while theory dictates a value of 6·45 cm.$^{-1}$†

* Epstein, *AP*, 1916, **50** 489, arrived at this result, but wrote his quantum numbers n_1, n_2, n_3; to convert these to the notation of this book, write $n_1 = (l - m_l)$, $n_2 = (n - l - 1)$, $n_3 = m_l$. Kramers, *ZP*, 1920, **3** 199, used the quantum numbers of this book, but wrote them n, n_1 and n_2; n is the chief quantum number, $n_1 = l$ and $n_2 = m_l$. Both neglect the spin of the electron so that $l = j$. For a recent account see Minkowski, *Hb. d. Phys.* 1929, **21** 393.

† Kassner, *ZP*, 1933, **81** 346.

Some of the hydrogen orbits dictated by this equation are calculated in Fig. 9·3 and shown diagrammatically in Fig. 9·4. The numbers of states having the same n, but different values of l or m_l, is $\frac{1}{2}n(n+1)$, but some coincide so that the number of energy levels is only $(2n-1)$.

n.	ns.	m_l
	12	0
	8	1
	4	2,0
4	0	3,1
	−4	2,0
	−8	1
	−12	0
	6	0
	3	1
3	0	2,0
	−3	1
	−6	0
	2	0
2	0	1
	−2	0
1	0	0

Fig. 9·4. Energy levels of a hydrogen atom in an electric field.

Passing from levels to lines, transitions are permitted only if

$$\Delta m_l = 0 \quad \text{or} \quad \pm 1.$$

When $\Delta m_l = 0$, the lines produced are polarised parallel to the field, while when $\Delta m_l = \pm 1$, they are polarised circularly in a plane normal to the field; the transitions with $\Delta m_l = 1$

give light circularly polarised in one sense, those with $\Delta m_l = -1$ in the opposite sense, but in the Stark effect the two oppositely polarised waves are always emitted together. The permitted transitions for the H_α line are calculated in Fig. 9·5 and shown diagrammatically in Fig. 9·6.

The intensities have been calculated by Schrödinger using the quantum mechanics,* and some measurements by Foster and Chalk† provide satisfactory confirmation of theory. The experi-

Polarisation	m_l		ns		$\Delta\nu$
	$n=3$	$n=2$	$n=3$	$n=2$	
	2	1	0	0	0
	1	0	3	2	1
				−2	5
			−3	2	−5
σ				−2	−1
	0	1	6	0	6
			0	0	0
			−6	0	−6
	1	1	3	0	3
			−3	0	−3
	0	0	6	2	4
π				−2	8
			0	2	−2
				−2	2
			−6	2	−8
				−2	−4

Fig. 9·5. Calculation of the Stark components of H_α.

mental difficulties however are great, for within a single Balmer line the intensity ratio of two components may be more than 1000 : 1, and ratios of this order cannot be measured; accordingly the authors worked only on components whose relative intensity was not more than 5 : 1; some of their results are shown in Fig. 9·7.

More recently Thornton‡ has worked with a trace of hydrogen in an inert gas, as this seems to offer the nearest approach to that ideal in which each atom would radiate undisturbed by

* Schrödinger, *AP*, 1926, **80** 457.
† Foster and Chalk, *PRS*, 1929, **123** 108.
‡ Thornton, *PRS*, 1935, **150** 259.

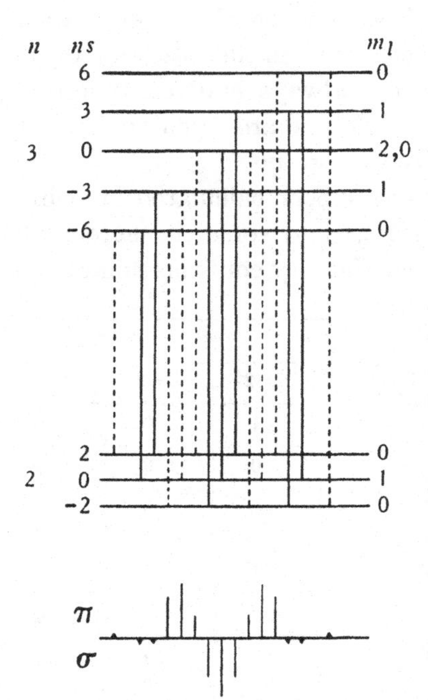

Fig. 9·6. Transitions producing H_α in an electric field. Transitions producing π components are dotted, those producing σ components drawn full.

Components compared		Measurements		Calculated by Schrödinger
		Stark[a]	Foster and Chalk	
H_α	π^{3}_{2}	1·10	3·32	3·16
	π^{3}_{4}	0·92	1·38	1·37
	σ^{0}_{1}	2·6	1·40	1·42
H_β	π^{8}_{10}	0·79	1·04	1·06
	π^{8}_{6}	1·90	4·59	4·74
	σ^{4}_{8}	1·30	1·56	1·56
	σ^{4}_{2}	3·80	6·23	6·35
H_γ	π^{18}_{15}	1·50	1·18	1·14
	π^{5}_{2}	0·94	1·03	1·23
	σ^{13}_{10}	1·42	0·93	0·94
	σ^{0}_{3}	2·25	1·71	3·07

Fig. 9·7. Intensity ratios of some components of three Balmer lines.

[a] Stark, *AP*, 1915, **48** 193.

other atoms. Small variations have been observed from the simple Epstein theory, but these are mainly due to the fine structure, which exists in zero field; the changes occurring as the field is increased were sketched by Schlapp* and these have been confirmed.

3. Elements other than hydrogen

The most commonly observed result of switching on an electric field is that lines appear which the selection rules forbid.† But in addition an electric field displaces the lines of the spectrum, and splits them into a number of components. In the Zeeman effect the individual components are displaced, but the pattern is symmetrical about the position of the undisplaced line; in the

Term	S	D			F			P	
M value of initial state	0	0	1	2	0	1	2	0	1
Polarisation of 4922 group	$\pi\sigma$	$\pi\sigma$	$\pi\sigma$	σ	$\pi\sigma$	$\pi\sigma$	σ	$\pi\sigma$	$\pi\sigma$
Polarisation of 3965 group	π	π	σ	—	π	σ	—	π	σ
Field in kv. cm.$^{-1}$: 0	+506·2		0			−5·6		−46·4	
10	506·3	2·0	1·75	1·0	−7·2	−7·05	−6·6	−46·9	−46·7
20	506·4	5·8	5·2	3·1	−9·6	−9·3	−8·7	−48·4	−47·9
40	507·0	14·9	13·3	7·9	−13·7	−13·6	−13·5	−54·0	−51·7
80	509·4	35·3	31·5	18·0	−18·2	−18·1	−23·6	−72·3	−65·4
100	512·4	45·3	41·1	23·2	−19·8	−19·2	−28·8	−83·5	−73·9

Fig. 9·8. Theoretical displacements of the 4922 and 3965 A. line groups of helium. These groups arise as $4\,^1L \rightarrow 2\,^1P$ and $4\,^1L \rightarrow 2\,^1S$; though the displacements are the same, the components which appear and their empirical polarisations are different.

Stark effect the pattern is not symmetrical and all the components are commonly displaced in the same direction, for the displacement is commonly much larger than the splitting; thus in many spectra low resolution reveals only the displacement. As in addition the laws which govern the displacement and splitting are different, the two phenomena are best treated separately.

To clarify these statements, consider the helium lines 4922 A. and 5047 A.,‡ which arise as $4\,^1D \rightarrow 2\,^1P$ and $4\,^1S \rightarrow 2\,^1P$. In an

* Schlapp, *PRS*, 1928, **119** 313.

† Koch first observed this, *AP*, 1915, **48** 98.

‡ Foster, *PRS*, 1927, **117** 145.

electric field two new lines appear nearby with wave-lengths, which show them to arise as $4\,^1P \rightarrow 2\,^1P$ and $4\,^1F \rightarrow 2\,^1P$ transitions, both previously forbidden. Thus in an electric field, we deal conveniently with a group of lines; this particular group is said to

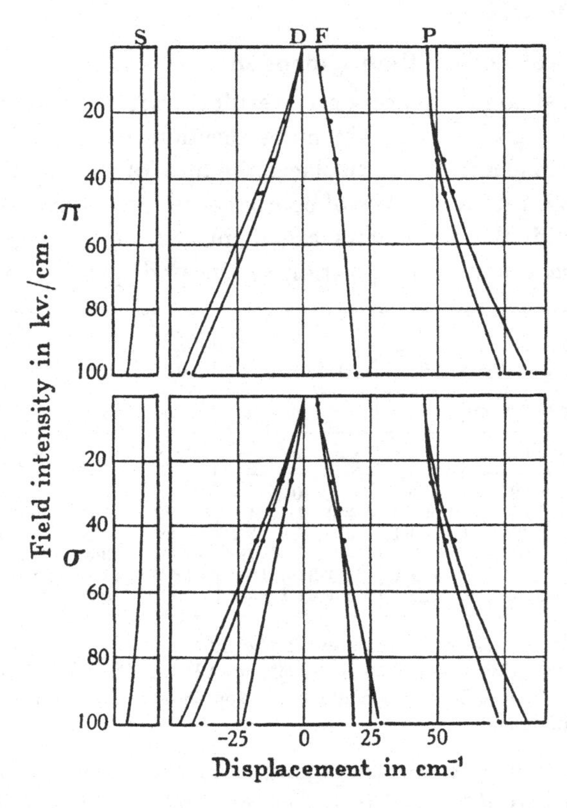

Fig. 9·9. Displacements of the 4922 A. line group of helium in various fields. The curves give the predictions of theory, the points are experimental. After Foster, *PRS*, 1927, **117** 160.

arise in the transition $4\,^1L \rightarrow 2\,^1P$, where L is an abbreviation for S, P, D or F. The displacements of the group are tabulated in Fig. 9·8 and shown graphically in Fig. 9·9, the π components appearing above and the σ components below. These displacements are derived from theory, but the experimental points inserted on the graph agree so closely with theory that the distinction is here unimportant. The splitting shown has its origin entirely in the upper level, for though the $2\,^1P$ term splits, theory shows that even in a field of 100,000 volt/cm. the interval is only 0·05 cm.$^{-1}$ This last is not a feature peculiar to the 4922 A. group;

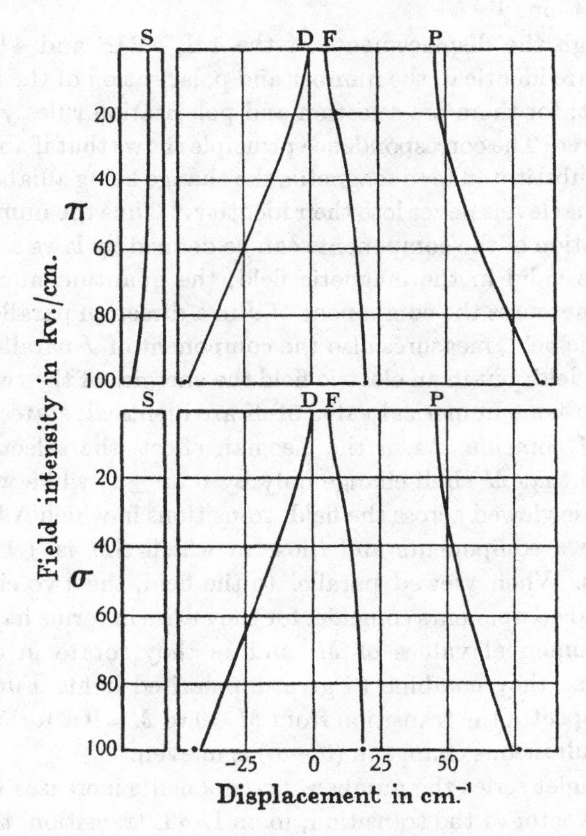

Fig. 9·10. Displacements of the 3965 A. line group of helium; the curves are theoretical, the points experimental. After Foster, *PRS*, 1927, **117** 160.

the electric field usually splits the upper levels more widely than the lower.

These observations can be confirmed by an examination of the 3965 A., $4\,^1P \rightarrow 2\,^1S$ line (Fig. 9·10), for in an electric field three new lines appear to complete the group $4\,^1L \rightarrow 2\,^1S$, and the displacements of these four lines are identical with those of the

4922 A. group. In this group the large displacements may well be contrasted with the small splitting intervals; in a field of 100,000 volt/cm., the $4\,^1P$ level is displaced by 30 cm.$^{-1}$ but split by only 10, and for the $4\,^1D$ level the corresponding figures are 40 and 4 cm.$^{-1}$

Though the displacements of the $4\,^1L \rightarrow 2\,^1P$ and $4\,^1L \rightarrow 2\,^1S$ groups are identical, the number and polarisation of the lines are different; for there are selection and polarisation rules yet to be considered. The correspondence principle shows that if an electric field is substituted for a magnetic, the change being adiabatic, the individual levels never lose their identity.* Thus the number and polarisation of the components can be defined by laws analogous to those valid in the magnetic field; the quantum number M, which measures the component of J in a direction parallel to the magnetic field, measures also the component of J parallel to the electric field; but in an electric field the energies of the two states with the same numerical value of M are identical; states defined by $\pm M$ coincide. As in the Zeeman effect, the selection rule requires that M shall change only by 0 or ± 1; while when the pattern is viewed across the field, transitions in which ΔM is zero produce π components and those in which ΔM is ± 1 σ components. When viewed parallel to the field, the two circularly polarised components coincide, for they arise in terms having the same numerical values of M, and as they rotate in opposite directions they combine to give unpolarised light. Further, in simple spectra the transition from $M=0$ to $M=0$ is forbidden, if the simultaneous change in $(J-L)$ is uneven.

In singlet series the number of components increases with the orbital vector of the transition; in an $L \rightarrow L$ transition, there are $(L+1)$ π components and $2L$ σ components; while when a term defined by L combines with one having a larger orbital vector, there are $(L+1)$ π components and $(2L+1)$ σ components. Level schemes designed for lines of the principal, sharp and diffuse series illustrate this statement (Fig. 9·11); while when the 4922 A. and 3965 A. groups are re-examined, the transitions observed are found to be just those predicted (Fig. 9·12).

* Kramers, *ZP*, 1920, **3** 199.

Though theory is so successful with singlets, there is still some doubt just how it should be applied to multiplet lines. In some P→S absorption lines of sodium* and potassium†, this theory is satisfactory; but in neon‡ Foster and Rowles have shown that the patterns displayed by triplet lines are often identical with those displayed by singlets, and that the former are never more com-

Series	Transitions	Number of components	
	M	π	σ
$^1S-{}^1S$	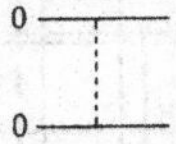 0 0	1	0
$^1P-{}^1S$ $^1D-{}^1S$ $^1F-{}^1S$	1 0 0	1	1
$^1P-{}^1P$	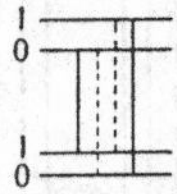 1 0 1 0	2	2
$^1D-{}^1P$ $^1F-{}^1P$ $^1G-{}^1P$	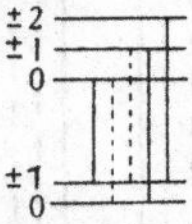 ±2 ±1 0 ±1 0	2	3

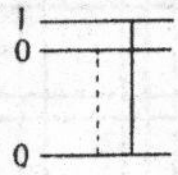

Fig. 9·11. Number and polarisation of the components into which the lines of various singlet series may be expected to split.

plicated; and this naturally suggests that in the theory M_L should be read for M_J. To meet this difficulty Foster and Rowles suggested that the field is really 'strong' in the technical sense, and that the splitting is therefore determined by M_L and M_S; if to this we add that M_S is so little affected by the electric field that in fact the complexities, which it should introduce, do not appear, the divergence between theory and experiment is avoided. In close analogy with the Zeeman effect, this theory commonly

* Ladenburg, *ZP*, 1924, **28** 51.

† Grotrian and Ramsauer, *PZ*, 1927, **28** 846.

‡ Foster and Rowles, *PRS*, 1929, **123** 81.

assumes that the (**LS**) coupling is broken when the multiplet interval is much smaller than the splitting due to the electric

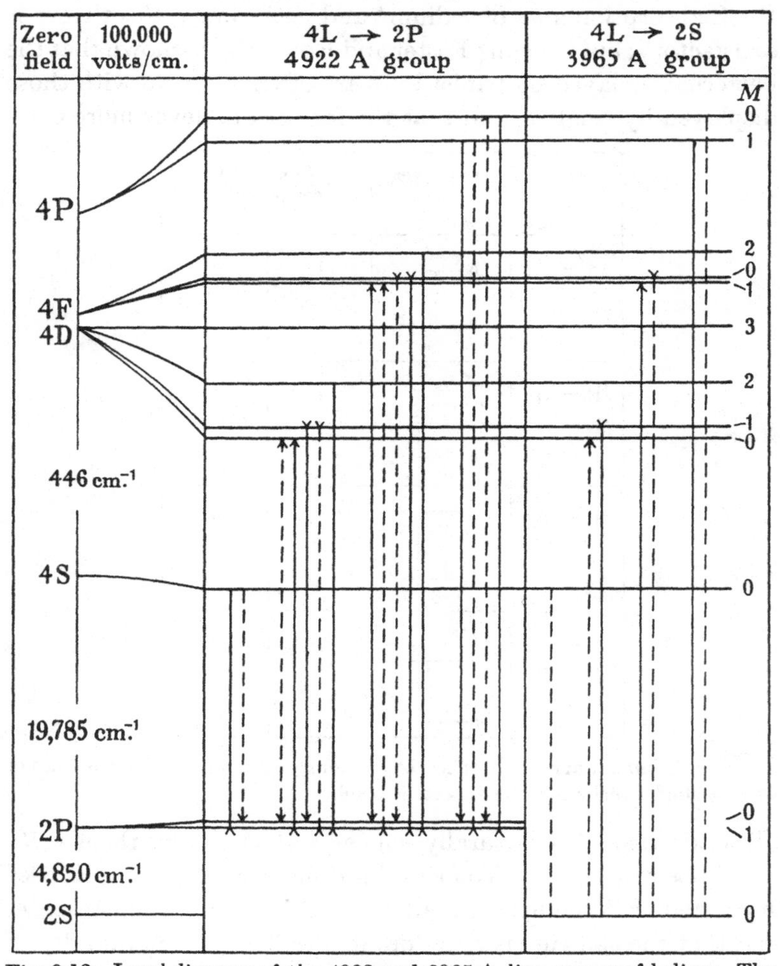

Fig. 9·12. Level diagram of the 4922 and 3965 A. line groups of helium. The π components are dotted, the σ components drawn full.

field. In their analysis of xenon, however, Harkness and Heard * found no support for this hypothesis.

In xenon, lines of the sharp, principal and fundamental series seldom yield more than one unpolarised component, showing that

* Harkness and Heard, *PRS*, 1933, **139** 416. For Kr: Ryde, *ZP*, 1933, **83** 354.

these states are not resolved by the field and spectroscopic dispersion used. On the other hand the diffuse lines are usually complex, and of the states producing them sixteen were examined in detail; in three p and eleven d states, the number of Stark levels is determined by M_J as theory requires, but in two d states the number is determined by M_L (Fig. 9·13). This would raise no difficulties if we could assume that in the former the (**LS**) coupling holds and in the latter breaks, but unfortunately the two states,

Wave-length	Transition	Change in J	Polar-isation	Displacement cm.$^{-1}$	Pattern determined by
6293	$6d_4 \to 2p_8$	$3 \to 3$	π	$-0·9$ $+0·2$ 3·2 6·6	
			σ	$-0·9$ $+0·2$ 3·2 6·6	
6182	$6d_4 \to 2p_9$	$3 \to 2$	π	$+0·2$ 3·5 7·0	M_J
			σ	$-0·9$ $+0·2$ 3·5 7·0	
5931	$6d_6 \to 2p_{10}$	$0 \to 1$	π	$+11·1$	
			σ	$+11·1$	
5895	$6d_5 \to 2p_{10}$	$1 \to 1$	π	$+12·5$ (d)	
			σ	$+4·1$ 12·5 (d)	
6473	$5d_2 \to 2p_{10}$	$1 \to 1$	π	$-0·1$ (d)	M_L
			σ	$-0·1$ (d) $+2·2$	
7266	$5d_2 \to 2p_7$	$1 \to 1$	π	$-0·1$ (d)	
			σ	$-0·1$ (d) $+2·2$	

Fig. 9·13. Examples of the splitting of the lines of xenon in an electric field. (d) Two components not resolved on the plate measured.

$5d_2$ and $6d_5$, whose levels are determined by M_L, differ in no systematic way from the other fourteen levels. Moreover, when the field increases from zero to 67000 volts/cm., no sign of a transition from one type to the other is observed. This result, combined as it is with a consistent failure to observe the complexities theoretically due to M_S, seems to rule out a Paschen-Back effect on the magnetic model, but what is to be put in its place is not yet clear.

Passing from the splitting to the displacement, the rule usually given to predict its magnitude is due to Bohr; * a term shows a large displacement when its quantum defect is small; this occurs

* Bohr, *PM*, 1914, **27** 506; *Proc. Phys. Soc.* 1923, **35** 300.

when the energy approximates to that of a hydrogen term, and the orbit does not penetrate the core. Thus in the alkalis a D term has a much smaller quantum defect than a P or S term, so that in the weak electric field, which is commonly produced by surrounding ions, lines arising in a D→P transition appear 'diffuse', while lines arising in an S→P transition appear 'sharp'.

This rule is satisfactory in many simple spectra, but in complex spectra it is often erroneous; indeed Pauli* and Condon† have shown that it is only a special case of a more general law. According to the wave mechanics, large displacements will occur when two terms, which are so related that the selection rules allow them to combine, have nearly the same energy. As low configurations seldom overlap, this explains why Stark displacements are confined to the higher levels; in Xe I not one of the 1s, 2p or 3d states,‡ the final states of all optical lines, is displaced, unless indeed all are displaced equally. Moreover, Condon has applied this theory to previously uncorrelated measurements in Ni I, Li I, C II and A I, and has shown that it describes them admirably, though of course the description is little more than qualitative.§

To illustrate this theory consider the rather complex spectrum of nickel, in which Takamine|| observed the Stark displacements a decade before Russell¶ conducted an analysis. Of the fifty lines measured, only some twelve are available for the present purpose, for Russell's analysis did not identify all the lines. The initial states of these lines are developed from the $3d^9\,4d$ or the $3d^8\,4s\,5s$ configurations, and lie between 49000 and 51000 cm.$^{-1}$ above the ground level; while the final states arise in the $3d^9\,4p$ or $3d^8\,4s\,4p$ configurations, and lie between 25000 and 33000 cm.$^{-1}$ At the latter height there is no even configuration, so that the whole of the observed displacement may be assumed due to the high levels, which are intermingled with odd terms from $3d^9\,5p$.

* Pauli, *Hb. d. Phys.* 1926, **23** 138. † Condon, *PR*, 1933, **43** 648.

‡ The notation here is Paschen's; see chapter XIV.

§ The agreement in Xe I is also satisfactory. Harkness and Heard, *PRS*, 1933, **139** 430.

|| Takamine, *AJ*, 1919, **50** 23. No splitting of the lines was observed.

¶ Russell, *PR*, 1929, **34** 821.

Wave-length A.	Transition	Field kv./cm.	Displacement cm.$^{-1}$ π	σ	Perturbed term	Nearest perturbing term	Interval cm.$^{-1}$
4410·50	$e\ ^3F_4 \rightarrow z\ ^5D_3^0$	39	+6·1	+4·1	$d^9.4d.e\ ^3F_4$	$d^9.5p.\ ^3D_3^0$	−5·2
4937·33	$e\ ^3F_4 \rightarrow z\ ^5F_4^0$	38·5	+5·0	+5·5			
5084·07	$e\ ^3F_4 \rightarrow z\ ^3D_3^0$	38·5	+5·0	+5·4			
5462·48	$e\ ^3F_4 \rightarrow z\ ^1F_3^0$	21·8	+2·9	+1·9			
5082·38	$e\ ^3P_1 \rightarrow z\ ^3P_1^0$	38·5	−1·7	−0·97	$d^9.4d.e\ ^3P_1$	$d^9.5p.\ ^3D_2^0$	+12·0
5184·59	$e\ ^3P_1 \rightarrow z\ ^3D_2^0$	21·8	−1·1	−1·1			
5587·85	$y\ ^3D_3^0 \rightarrow a\ ^3P_2$	21·8	+0·48	+0·35	$d^8.s.4p.y\ ^3D_3^0$	$d^9.5s.\ ^3D_3$	−15·11
5018·30	$e\ ^3F_2 \rightarrow z\ ^3D_1^0$	38·5	−2·8	0·0	$d^9.4d.e\ ^3F_2$	$d^9.5p.\ ^3D_1^0$	+16·7
5155·76	$e\ ^1F_3 \rightarrow z\ ^1D_2^0$	21·8	+3·0	+3·0	$d^9.4d.e\ ^1F_3$	$d^9.5p.\ ^3F_4^0$	−42·5
5146·48	$e\ ^5F_4 \rightarrow z\ ^3D_2^0$	21·8	+0·44	+0·26	$d^8.s.5s.e\ ^5F_4$	$6^0\ (J=3)$	−53·3
5176·56	$f\ ^1D_2 \rightarrow z\ ^1D_2^0$	21·8	+0·56	+0·34	$d^9.4d.f\ ^1D_2$	$d^9.5p.\ ^1D_2^0$	−65·0
5142·77	$f\ ^3D_2 \rightarrow z\ ^3D_2^0$ $f\ ^3D_3 \rightarrow z\ ^5F_3^0$	21·8	+0·34	+0·26	$d^9.4d.f\ ^3D_2$ $d^9.4d.f\ ^3D_3$	$d^9.5p.\ ^3D_3^0$ $d^9.5p.\ ^3D_2^0$ $d^9.5p.\ ^3D_3^0$	− 0·5 −86·5 +56·1

Fig. 9·14. Displacements of some nickel lines in an electric field. The first two columns give the wave-length and the transition in which it arises; next appear the fields employed and the displacements observed by Takamine; the last three columns contain the high term of the transition, the term which perturbs it and the interval separating the two. The lines have been ordered by the interval in the last column, no attention being paid to sign. Condon, *PR*, 1933, **43** 651.

Fig. 9·14 provides satisfactory confirmation of two predictions. Of the lines available only three are displaced to lower wave-numbers, and all three arise from terms which have a perturbing term lying close above them in the energy scale. Secondly, having arranged the lines so that the interval between perturbing term and perturbed increases, the displacement shows a general tendency to diminish; the sequence is not very regular, but with the exception of the 5156 A. line, it is perhaps as good as can be expected.

At the bottom of the table is added a line, 5142 A., to which Russell gave two alternative assignments. If in fact it arises from the f^3D_2 term, the theory of the Stark effect would suggest a strong upward perturbation, whereas if it arises from f^3D_3, the perturbation would probably be small, since the nearest terms are distant and there is one on either side. As the actual Stark displacement of 5142 A. is quite small, there seems no doubt that the second alternative is correct.

Condon claims that this is the first time a doubtful point in the analysis of a complex spectrum has been settled by use of an electric field, a fact which perhaps explains why so few people ever read a chapter on the Stark effect.

BIBLIOGRAPHY

The best critical account of theory and experiment is by Minkowski in *Handbuch der Physik*, 1929, **21** 389–439. Stark has given a rather more thorough account of the experimental work in *Handbuch der Experimental Physik*, 1927, **21** 399–562. Harkness and Heard in *PRS*, 1933, **139** 416, give a valuable discussion of more recent problems.

CHAPTER X

THE PERIODIC SYSTEM

1. The table

If the elements are arranged in order of increasing atomic number, and the atomic volume plotted, a periodic curve is obtained (Fig. 10·1); and if for the atomic volume one substitutes the melting point, the coefficient of expansion, the atomic refraction

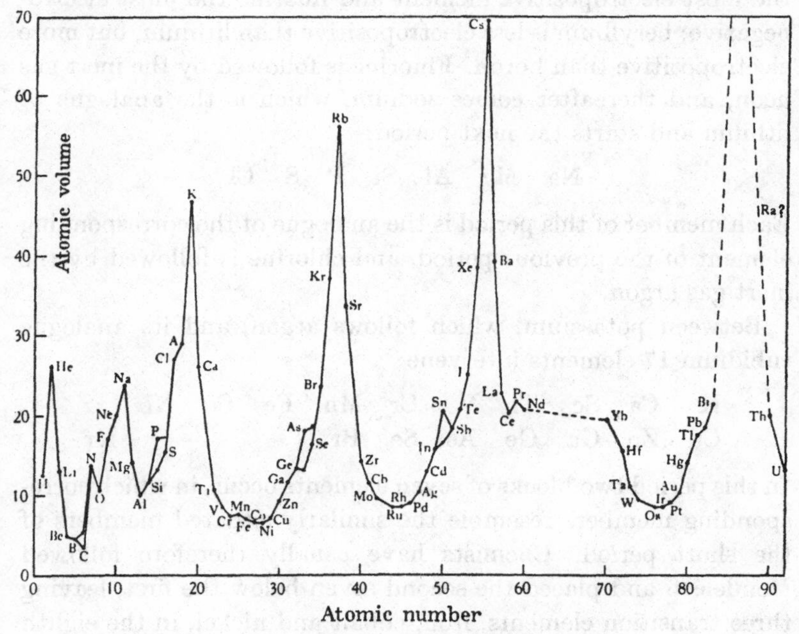

Fig. 10·1. Atomic volumes.

or the electrical conductivity, the periodicity remains. It appears too in chemistry; lithium resembles sodium more than any intervening element, and the same is true of carbon and silicon, and of fluorine and chlorine; thus if eight steps are taken from a given element, the element reached resembles that from which the start was made. This 'law of octaves' was stated by Newlands in

1864; to-day it needs revision only in that 18 or 32 must be substituted for eight in the later periods. Certain properties thus suggest that the 92 elements may be divided into seven periods, so chosen that each of the first six ends in an inert gas. This gives the table in the form shown in Fig. 10·2.

Within the periods the properties are regularly graded. Lithium is the first of seven elements:

Li Be B C N O F

in which the change is from metallic to non-metallic. Lithium is the most electropositive element and fluorine the most electronegative; beryllium is less electropositive than lithium, but more electropositive than boron. Fluorine is followed by the inert gas neon, and thereafter comes sodium, which is the analogue of lithium and starts the next period:

Na Mg Al Si P S Cl

Each member of this period is the analogue of the corresponding element of the previous period, and chlorine is followed by the inert gas argon.

Between potassium, which follows argon, and its analogue rubidium 17 elements intervene:

K	Ca	Sc	Ti	V	Cr	Mn	Fe	Co	Ni	
Cu	Zn	Ga	Ge	As	Se	Br				Kr

In this period two blocks of seven elements occur, in which corresponding members resemble the similarly situated members of the short period. Chemists have usually therefore followed Mendeléeff and placed the second seven below the first, leaving three transition elements, iron, cobalt and nickel, in the eighth column (Fig. 10·3). (I have used the word 'column' instead of the more usual 'group', because I have had to use 'group' in so many other senses.) Some of the similarities, however, are much less striking than in the earlier periods; potassium is as like sodium as sodium is like lithium, but almost the only link between these elements and copper is that on occasion all are monovalent. And though bromine is closely similar to chlorine, manganese is related only by a few heptavalent compounds. Accordingly

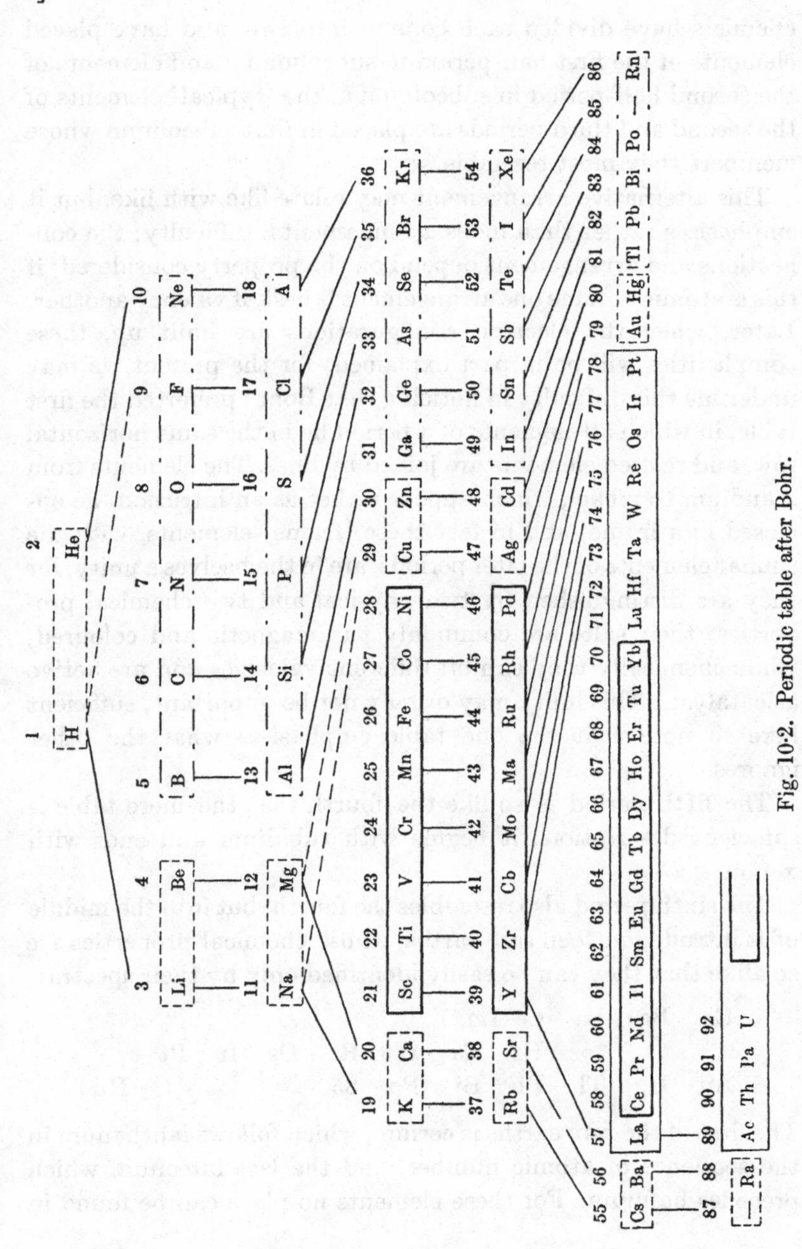

Fig. 10·2. Periodic table after Bohr.

chemists have divided each column into two, and have placed elements of the first half-period in subcolumn *a* and elements of the second half-period in subcolumn *b*; the 'typical' elements of the second and third periods are placed in that subcolumn whose members they most resemble.

This alternative arrangement may relate like with like, but it emphasises rather than hides a fundamental difficulty; the connections and arrangement depend on the property considered; if this is atomic volume one arrangement is best, if valency another. Later, when the electron configurations are built up, these complexities will be in part explained; for the present we may underline this difficulty in noticing that Bohr* preferred the first table, in which all elements of a period lie in the same horizontal row, and related elements are joined by lines. The elements from scandium to nickel, which appear rather as an intrusion, he enclosed in a frame; and in fact these 'frame' elements, with the similar elements of the later periods, are in themselves a unity, for they are distinguished by two physical and two chemical properties; their salts are commonly paramagnetic and coloured; while chemically they exhibit different valencies and are active as catalysts. This unity may or may not be important; sufficient here to note that the one table emphasises what the other ignores.

The fifth period is so like the fourth that the mere table is sufficient description; it begins with rubidium and ends with xenon.

The sixth period also resembles the fourth, but into the middle of it intrude fourteen rare earths, whose chemical properties are so alike that they can be easily identified only by their spectra:

Cs	Ba	La	Ce–Lu							
			Hf	Ta	W	Re	Os	Ir	Pt	
Au	Hg	Tl	Pb	Bi	Po	85				Rn

The first of the rare earths is cerium, which follows lanthanum in the sequence of atomic number, and the last lutecium, which precedes hafnium. For these elements no place can be found in

* Bohr, *Theory of spectra and atomic constitution*, 1922, 70.

Period	Column I *a*	Column I *b*	Column II *a*	Column II *b*	Column III *a*	Column III *b*	Column IV *a*	Column IV *b*	Column V *a*	Column V *b*	Column VI *a*	Column VI *b*	Column VII *a*	Column VII *b*	Column VIII *a*	Column VIII *b*
I	1H															2He
II	3Li		4Be			5B		6C		7N		8O		9F		10Ne
III	11Na		12Mg			13Al		14Si		15P		16S		17Cl		18A
IV	19K	29Cu	20Ca	30Zn	21Sc	31Ga	22Ti	32Ge	23V	33As	24Cr	34Se	25Mn	35Br	26Fe 27Co 28Ni	36Kr
V	37Rb	47Ag	38Sr	48Cd	39Y	49In	40Zr	50Sn	41Cb	51Sb	42Mo	52Te	43Ma	53I	44Ru 45Rh 46Pd	54Xe
VI	55Cs	79Au	56Ba	80Hg	57La + ΣCe	81Tl	72Hf	82Pb	73Ta	83Bi	74W	84Po	75Re	85—	76Os 77Ir 78Pt	86Rn
VII	87—		88Ra		89Ac		90Th		91Pa		92U					

ΣCe represents:	58	59	60	61	62	63	64	65	66	67	68	69	70	71
	Ce	Pr	Nd	Il	Sm	Eu	Gd	Tb	Dy	Ho	Er	Tu	Yb	Lu

Fig. 10·3. Periodic table after Mendeléeff.

the Mendeléeff table, at least if the regularities of the table are to be preserved; even the transition elements, occurring as they do three in one column, are a serious difficulty.

Within the columns only similarities have thus far been emphasised, but in fact the properties vary regularly with the atomic number. In general the elements grow more metallic as the atomic number increases; lead is more metallic than carbon, and bismuth than nitrogen, but subcolumns I*b* and II*b* seem to be exceptions to this rule.

2. Valency

Valency as originally introduced is merely a number. The valency of an element is the number of monovalent atoms with which an atom of the element combines, a monovalent atom being at first hydrogen or chlorine. Without going further than this simple definition, valency is seen to be a periodic function of the atomic number, for typical compounds of successive columns of the system are

NaH	CaH_2	B_2H_6	CH_4	NH_3	H_2S	HCl	
KCl	$MgCl_2$	$AlCl_3$	$SiCl_4$	PCl_5	WCl_6		$PdCl_4$
K_2O	CaO	Al_2O_3	CO_2	P_2O_5	SO_3	Cl_2O_7	OsO_4

Thus the hydrogen valency rises from one to four in the first half of the period and then sinks again to one, while the chlorine and oxygen valencies rise steadily from one to eight in successive columns; though in the elements which succeed phosphorus lower valencies are known. Osmium forms OsF_8 though no octachloride is known.

The hydride of boron has the formula B_2H_6, when we might expect BH_3, and closer examination reveals a host of such petty irregularities; but these excrescences do not hide the essential simplicity of the scheme. When, however, one no longer remains content with the question, 'How many atoms combine?' and asks instead 'How do atoms combine?' the petty irregularities, so characteristic of chemistry, obtrude themselves provokingly. In recent years, however, chemists have found that there are two types of compound, which differ so profoundly that one may suspect that the mechanism which holds the atoms together is different. These two types are known as*

(1) Electrovalent compounds, which ionise in water and are formed by the electrostatic attraction of ions.

(2) Covalent compounds, which do not ionise in solution; organic compounds are typically covalent.

The properties by which these two types may be recognised must be mastered before their structure can be satisfactorily explained.† A covalent bond is a definite directed link extended as a man might extend a hand to a friend, whereas electrovalency is a vague general attraction like that exerted by a Hyde Park orator over his crowd. From this essential difference three consequences arise.

Covalency is directed, electrovalency is not; so that only covalent compounds can possess isomers or stereo-isomers.

A covalent bond is not broken in solution, whereas a solvent of high dielectric constant will greatly reduce the force between two electrostatic charges. The most satisfactory test here is electrolysis, but any method which can be used to measure molecular weight, such as the cryoscopic, can be used also to discover whether a compound dissociates. Again one may test for a single ion, as by adding an alkaline halide to a solution suspected of containing silver ions; for example, the soluble silver thiosulphate $Na[S_2O_3Ag]$ gives a precipitate with sodium iodide, but not with sodium chloride, showing that the silver is almost wholly in a covalent state, but that a minute quantity of the complex ion dissociates liberating silver ions.

* Sidgwick, *Electronic theory of valency*, 1929, 52f; Sidgwick, *The covalent link in chemistry*, 1933, 61. Many chemists, however, think that electrovalent and covalent compounds are the extremes of a continuous range; sharp division, they say, is no more possible here than between yellow and green in the spectrum.

† Van Arkel and de Boer in *Chemische Bindung als electrostatische Erscheinung* (German translation from the Dutch by L. and W. Klemm, 1931) treat all compounds as electrovalent, and explain the characteristic covalent properties as occurring when the central ion is surrounded. This draws attention to certain interesting relations between the size of the ion and the valency type, but it is difficult to see how two ions can 'surround' a third however large they are, yet $BeCl_2$ is covalent; and the contrasting properties are not satisfactorily explained. The authors make no reference to Sidgwick's book.

A covalent bond again usually so satisfies both atoms that there is no external or stray field, whereas when two ions unite a stray electric field exists round the pair. By an appropriate 'head to tail' arrangement the molecules of an ionised compound can reduce this stray field, but the work required to separate the molecules is greater in consequence. Accordingly, an ionised compound has to be raised to a much higher temperature than a covalent one before it boils, and in general it dissolves but little in an inactive solvent, such as benzene. These properties divide the hydrides into two distinct groups. Methane, CH_4, and hydrogen chloride are volatile and hardly conduct when pure; they are therefore covalent compounds. Sodium hydride and calcium hydride, on the other hand, are non-volatile, and are good conductors when fused, hydrogen being liberated at the anode on electrolysis; they are therefore salts containing a negative hydrogen ion. The halides too can be divided into volatile non-salts and non-volatile salts, and in these two classes the effect of increasing the atomic weight of the halogen is different. Silicon fluoride and sodium fluoride are typical.

Boiling points in ° C. *at* 760 *mm. pressure*

NaF	NaCl	NaBr	NaI
1695°	1441°	1393°	1300°
SiF_4	$SiCl_4$	$SiBr_4$	SiI_4
− 90°	57°	153°	*c.* 290°

In normal substances the boiling point rises with the molecular weight, and in the non-polar halides this actually happens; but in the salts the electrostatic attraction of the charged ions is the determining factor, and as the charges are the same the force must diminish as the atomic diameter of the halogen increases. The absolute boiling points of the sodium salts are roughly in inverse ratio to the distances between the atomic centres as determined from X-ray measurements.

In solution similar considerations arise; the mutual attraction between the ions of a solid tends to make it as little soluble as volatile; but a solvent of high dielectric constant like water diminishes the forces between the ions and may also promote solution by combining with the ions. This explains why salts, whose vapour pressure at room temperature is infinitesimal, may yet be very soluble in water; sodium chloride, for example, does not boil until 1441° C. A solvent, however, which is both saturated and of low dielectric constant, such as paraffin or benzene, has but slight chemical or physical effect, so that it cannot overcome the attraction of the ions and salts do not dissolve.

These criteria suffice to distinguish many simple compounds as either ionic or covalent. How are these two types to be explained in atomic theory?

In the ionic type Kossel* pointed out that the electropositive partner is commonly situate soon after an inert gas, while the electronegative partner is situate before; further, the valency is equal to the number of places the element is away from the gas; an alkaline earth is electropositive and divalent, a halogen electronegative and monovalent. These facts Kossel explained by supposing that the configuration of an inert gas is peculiarly stable, and near elements strive to attain it by acquiring or relinquishing electrons. For example, in the sequence

O	F	Ne	Na	Mg
8	9	10	11	12

ten electrons must be a particularly stable configuration, which fluorine assumes when it adds one electron, and magnesium when it loses two. Theory thus accounts for the stable ions O^{--}, F^-, Na^+ and Mg^{++}, which when they combine yield such neutral salts as NaF, MgO and MgF_2, the ions in the salt being held together by electrostatic forces. On this theory of valency, the formation of the ions, which is the interesting step, is essentially a problem in the mechanics of the atom; their subsequent attraction is the most elementary electrostatics.

The theory is in complete agreement with work in electrolysis, which shows that in solution ions do actually carry charges in the ratio postulated, and it also explains the results of experiments in which X-rays have been used to determine the structure of

* Kossel, *AP*, 1916, **49** 229.

crystals. In the alkaline halides, for example, two types of lattice occur; in the first, known as the rock-salt lattice, each sodium atom is surrounded symmetrically by six chlorine atoms, and each chlorine by six sodiums; the second, or caesium-chloride lattice, is similar except that the number is eight instead of six. These structures are close-packed systems in which there is no sign that a particular sodium atom is not equally attracted by all six chlorine atoms; in short, the structures are what might be expected if the only attraction is electrostatic.

In contrast the accepted theory of covalency assumes that a pair of electrons pass round two nuclei; but in putting forward this theory G. N. Lewis* assumed, like Kossel, that the cause of chemical combination is the tendency of the electrons to redistribute themselves among the atoms so as to form more stable configurations; an electron which passes round two nuclei may, Lewis assumed, help to complete stable groups in both. Fluorine, for example, readily takes up an electron to form a fluorine ion, but it can also complete its outer shell of eight by sharing one belonging, say, to another fluorine atom; and if at the same time the second atom shares an electron belonging to the first, both atoms are satisfied. The various states of the atom are shown below, the dots and crosses representing valency electrons, without any assumption as to their position in space:

```
                      ..
Fluorine atom        :F·
                      ..
                              ..
Fluorine ion                 :F·×
                              ..
                      ..     ××
Fluorine molecule    :F·×    F×
                      ..     ××
```

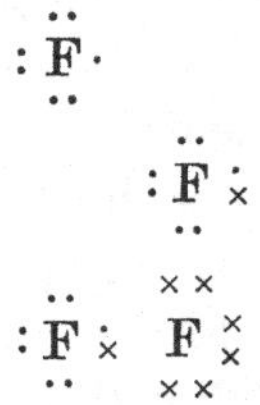

Two consequences of this theory are at once apparent; as it enables a number of electrons to be as stable as if there were more, it is likely to appear chiefly in atoms each of which is a few electrons short of a stable number. This agrees with experience, covalent compounds being far commoner among the electronegative elements than among the metals.

* Lewis, *Am. Chem. Soc. J.* 1916, **38** 762.

Again, the numerical value of the valency is the same on both theories. An electronegative element gains an electron for each covalency it forms, so that the covalency must be equal to the number of places it is away from an inert gas; but this is the electrovalency too. An electropositive atom on the other hand has not only to accept a share of a strange electron, but must give a share in one of its own, so that the number of bonds it can form is limited to the number of electrons it has to offer, and this again is the electrovalency. This explains why chlorine is monovalent in carbon tetrachloride as well as in sodium chloride,

$$\begin{array}{ccc} & \ddot{:Cl:} & \\ :\ddot{Cl} \times & \overset{\times\cdot}{C} & \times \ddot{Cl}: \\ & \underset{\cdot\cdot}{:Cl:} & \end{array} \qquad Na\left[\times \ddot{\underset{\cdot\cdot}{Cl}}:\right]$$

and why zinc is divalent in zinc methide as well as in zinc fluoride.

$$\begin{array}{ccc} H & & H \\ H \times \overset{\times\cdot}{\underset{\cdot\times}{C}} \times Zn \times & & \overset{\times\cdot}{\underset{\cdot\times}{C}} \times H \\ H & & H \end{array} \qquad \left[:\ddot{\underset{\cdot\cdot}{F}} \times\right] Zn \left[\times \ddot{\underset{\cdot\cdot}{F}}:\right]$$

Both theories of valency assume that atoms in combining tend to provide themselves with stable configurations. In covalent compounds these configurations are but little understood; the quantum mechanics asserts that the two shared electrons are equivalent, though they belonged originally to different atoms; but whether the number of electrons alone is significant, the shared electrons being counted with the rest, is not yet beyond dispute. Whatever the solution, however, the problem is one of molecular rather than atomic physics.

In contrast, the stable configurations of an ion concern only a single atom. The most serious difficulty indeed is to compile a list of ions, which is reasonably full and yet does not contain such imaginaries as C^{4+}, which have no existence in nature, the compounds in which they are supposed to occur being covalent. In the specialist books the two types seem always to be listed together, so that Fig. 10·4 has had to be compiled by a direct appeal to melting points, solubilities and the other criteria already cited. The table is certainly incomplete, but even so it reveals the stability of some configurations besides the inert gas type. Atoms which succeed cobalt, palladium and platinum tend to lose electrons and to assume their configurations. In the fifth and sixth periods, too, the configurations of cadmium and mercury seem more than usually stable; thallium is monovalent, tin and lead divalent, and antimony and bismuth trivalent. Finally, in the rare earths, the divalency of europium and the tetravalency of terbium may both be ascribed to the stability of a shell containing seven f electrons. This would be an unreasonable assumption if the p and d shells did not show a similar stability when just half full; the ionisation potentials of the short periods and the relative energies of the low configurations of the long periods leave no doubt of this stability, though it does not seem to manifest itself chemically.

3. Covalency

In the past the mechanism of chemical combination and all discussion of valency were the concern of the chemist alone; but in recent years much has been learnt from the study of band spectra, while the mathematician has shown that at least one general rule can be derived from the quantum mechanics. In the absence of any satisfactory synthesis, each of these three fields has its own literature.

The best balanced books seem to be *The electronic theory of valency*, 1927, and *The covalent link in chemistry*, 1933, both by N. V. Sidgwick; Sidgwick does not ignore the evidence from band spectra and the quantum mechanics, but his interests are primarily those of the chemist. *Chemische Bindung als electrostatische Erscheinung*, 1931, by van Arkel and de Boer, also contains much that is valuable; that these authors can attempt to show that all compounds are held together by electrostatic attraction, while the spectroscopist and mathematician are considering only shared electrons, shows however how far the different fields are apart.

The work on band spectra is more recent and therefore less

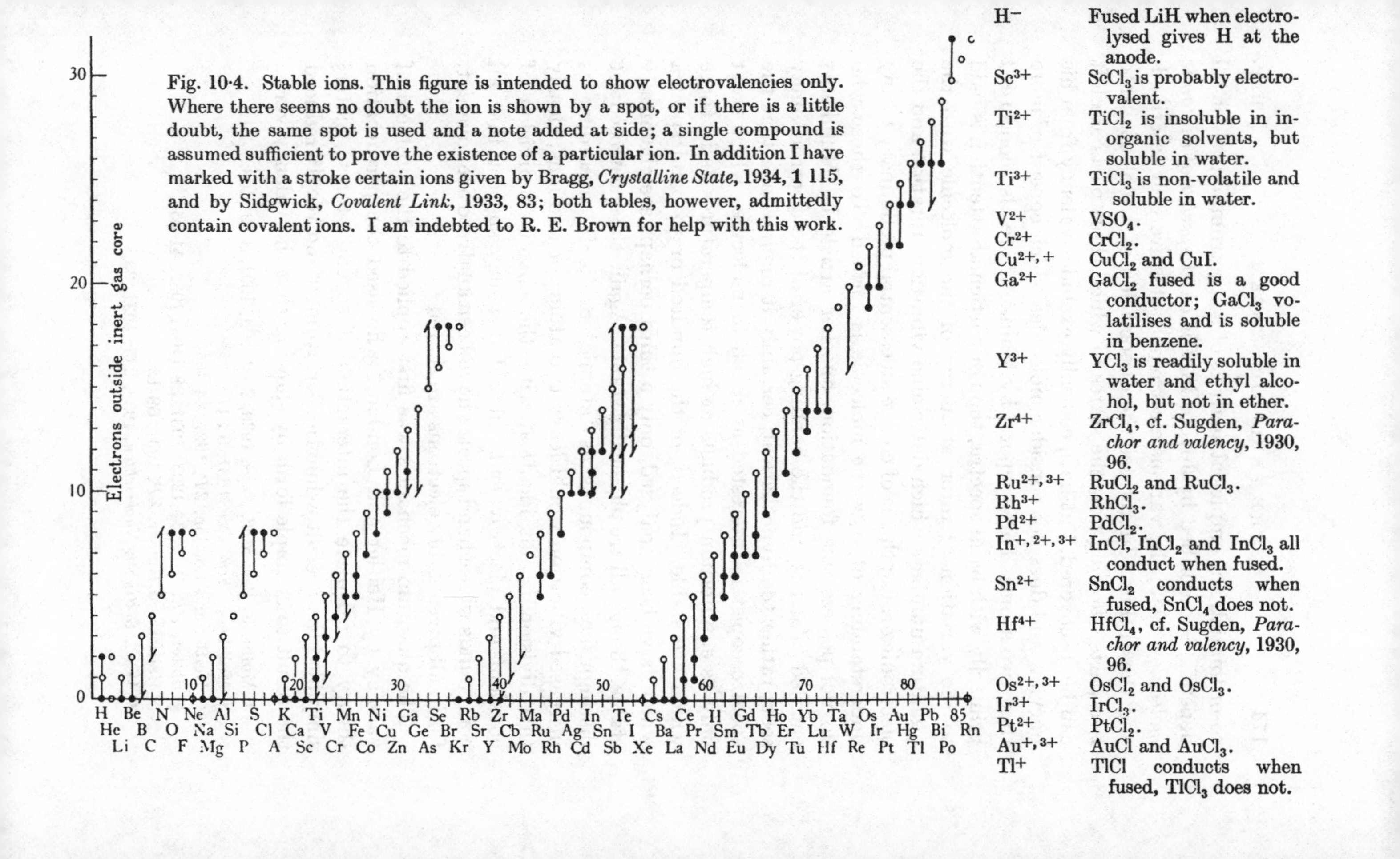

Fig. 10·4. Stable ions. This figure is intended to show electrovalencies only. Where there seems no doubt the ion is shown by a spot, or if there is a little doubt, the same spot is used and a note added at side; a single compound is assumed sufficient to prove the existence of a particular ion. In addition I have marked with a stroke certain ions given by Bragg, *Crystalline State*, 1934, 1 115, and by Sidgwick, *Covalent Link*, 1933, 83; both tables, however, admittedly contain covalent ions. I am indebted to R. E. Brown for help with this work.

co-ordinated; the empirical analysis of a spectrum, on which all depends, is described by Jevons in *The band spectra of diatomic molecules*, 1932; the various electronic states are well described, and the author adumbrates the dissociation of a molecule into two parts, showing how the states in which these parts are left can be discovered, but he specifically excludes valency from his purview and does not consider molecules which consist of more than two atoms. Three reports by Mulliken* are also concerned primarily with band spectra, though particular attention is paid to the variation of internal energy of the molecule with the quantum numbers which determine vibration, rotation and the electronic orbits; this is of course an essential preliminary to any understanding of how the molecule is formed. To those who do not possess this foundation, another article by Mulliken,† entitled 'Valency and the bonding power of the electron', may seem rather too hypothetical; certainly it reminds one that the spectroscopist is interested as much in molecules which exist only as equilibrium products at high temperatures as in those which are stable. Indeed, of the hundred or so band spectra which have been analysed, only a third perhaps are chemically stable, though all are physically stable. Again, the spectroscopist distinguishes compounds as 'atomic' or 'ionic' according as increased vibration would leave two atoms or two ions; clearly this division is not identical with the chemist's division of covalent and electrovalent; indeed one may suspect that all compounds whose band spectra have been analysed are covalent, for in all presumably electrons are shared.

The quantum mechanics was first applied to the problem of valency by Heitler and London,‡ who used the perturbation theory to calculate the interaction of several atoms. Slater§ simplified their result, while Heitler, Rumer|| and Weyl¶ reduced the result to a simple form by showing that the valency varies

* Mulliken, *Rev. Mod. Phys.* 1930, **2** 60, 506; 1931, **3** 89; 1932, **4** 1.

† Mulliken, *Chem. Rev.* 1931, **9** 347.

‡ Heitler and London, *ZP*, 1927, **44** 455.

§ Slater, *PR*, 1929, **34** 1293; 1931, **38** 1109; 1932, **41** 255.

|| Heitler and Rumer, *ZP*, 1931, **68** 12.

¶ Weyl, *Göttingen Nachrichten*, 1930, 285; 1931, 33.

with the multiplicity; when the multiplicity of the ground term is $(2S+1)$, the valency is $2S$. In this form the theory has been summarised by Born,* but unfortunately the author does little to render his theory palatable to the chemist. The statement that the valencies of the second period should be

Li	Be	B	C	N	O	F	Ne
1	0	1	2	3	2	1	0

is relegated to a footnote, and no attempt is made to develop this thesis in the later periods. Possibly all beryllium compounds are electrovalent, and possibly in carbon there is a low quintet term unidentified, which explains its usual quadrivalency, but so many like difficulties arise that much of inorganic chemistry will need reconsideration if the theory is to stand. This might be attempted in a book; it is impossible in a few pages.

4. The displacement law

If the arc spectrum of an element is of even multiplicity, the arc spectra of the elements which precede and succeed it in the periodic table will be odd. This rule, which is valid for spark

Column	I	II	III	IV	V	VI	VII	VIII			I	II	III
Element	K	Ca	Sc	Ti	V	Cr	Mn	Fe	Co	Ni	Cu	Zn	Ga
Atomic number	19	20	21	22	23	24	25	26	27	28	29	30	31
Arc spectrum	2	1 3	2 4	1 3 5	2 4 6	(1) 3 5 7	(2) 4 6 8	(1) 3 5 7	2 4 6	1 3 5	2 4	1 3	2
Spark spectrum	1 3	2	1 3	2 4	(1) 3 5	2 4 6	(1) (3) 5 7	(2) 4 6	(1) 3 5	2 4	1 3	2	1 3

Fig. 10·5. Multiplicities of the arc and spark spectra of the iron row; multiplicities believed to occur, but not yet identified, are enclosed in brackets.

* Born, *EEN*, 1931, 10 387.

spectra too, as Fig. 10·5 shows, is sometimes known as the alternation law. But if the arc spectrum of an element is of even multiplicity the spark spectrum is odd, so that if the spectrum of hydrogen is known to be doublet in structure, this law may be generalised to read: the multiplicity of a spectrum is even if it arises from an odd number of electrons, and odd if it arises from an even number.

The connection between a spectrum and the number of electrons from which it arises is, however, much closer than the alternation law alone suggests. The arc spectrum of sodium is similar to the spark spectrum of magnesium and the second spark spectrum of aluminium in many ways besides exhibiting the same multiplicity, and this is true of all isoelectronic spectra.* Stated thus the displacement law is simple enough, but in what precisely does this similarity consist?

The term systems of these three spectra, spectra which are commonly written Na I, Mg II and Al III, are shown in Fig. 10·6; but in order to fit them into a page each is drawn to a different energy scale, for just as a term of He II has four times the energy of the homologous term of H I because the nucleus carries twice the charge, so the energies of Mg II and Al III are roughly four and nine times the energies of Na I. In the spectra themselves this means that a visible line of Na I will have its Mg II and Al III homologues in the ultra-violet; thus the wave-length of the $2\,{}^2P_{1\frac{1}{2}} \rightarrow 1\,{}^2S_{\frac{1}{2}}$ line, which is the D_2 line of Na I, is 2796 A. and 1855 A. in Mg II and Al III respectively. If the energies change, however, the terms which occur and their relative positions remain roughly the same, as Fig. 10·6 shows clearly enough without explanation.

If other series of isoelectronic spectra are examined, all are found to have a common term system, and all exhibit the same change of energy scale in passing from the arc to the spark spectrum; but the relative energies of the terms do sometimes change; when this occurs among the high terms, it is obviously of less account than when the ground term changes, since the latter shows the normal state of the atom or ion. The sequence K I,

* Kossel and Sommerfeld, *Deutschen Phys. Ges.* 1919, **21** 240.

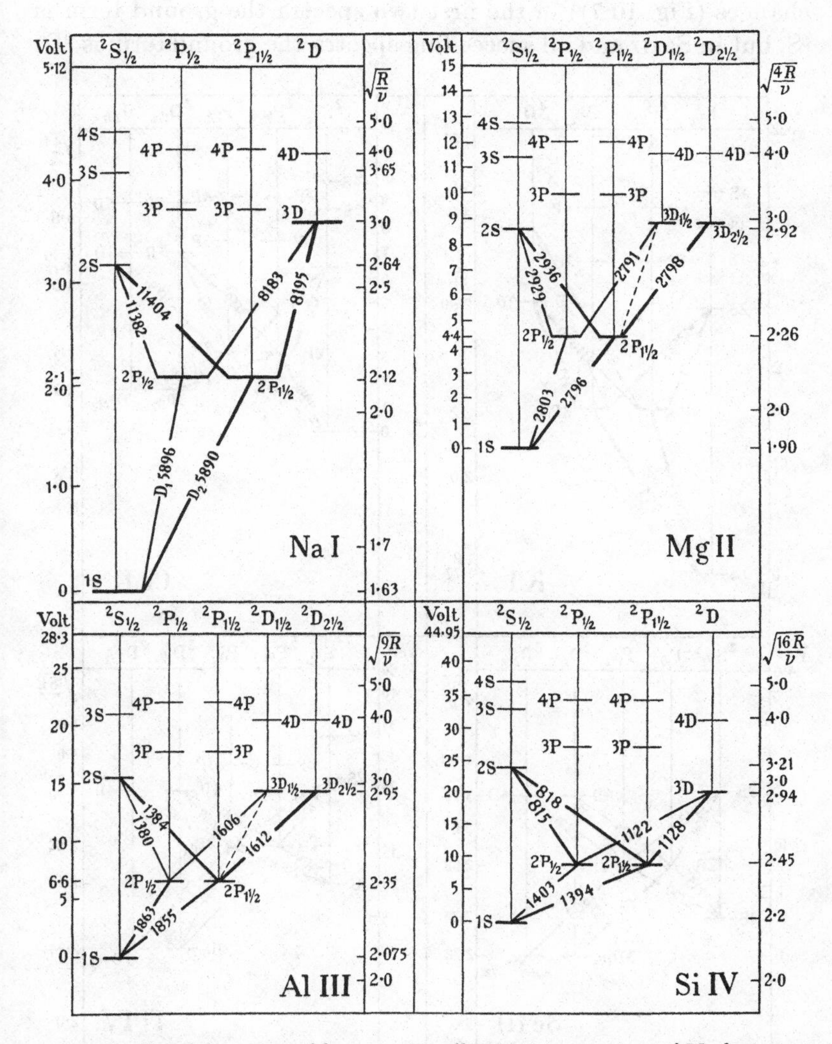

Fig. 10·6. Level diagrams of four spectra all arising in a system of 11 electrons, but each with a different nuclear charge. The D orbits sink relative to the S as the nuclear charge increases.

Ca II, Sc III, Ti IV illustrates one of the more striking of these changes (Fig. 10·7); in the first two spectra the ground term is ^2S, but in Sc III and all succeeding spectra the ground term is ^2D.

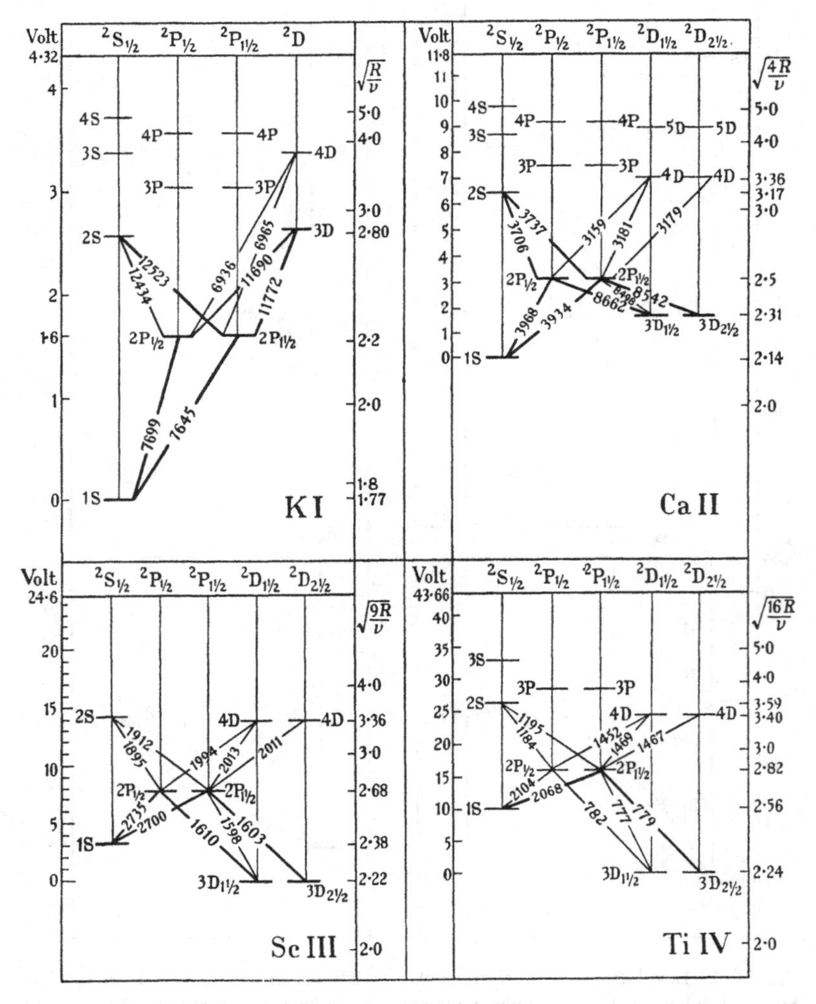

Fig. 10·7. Level diagrams of four spectra all arising in a system of 19 electrons, but each with a different nuclear charge. The ground term changes from ^2S to ^2D as the nuclear charge increases from 20 to 21.

5. Electronic structures

An explanation of the periodic system naturally assumes that the electrons may be divided into distinct groups, so that the grouping of the elements in the system is attributed to the gradual formation of groups of electrons within the atom. The grouping of the elements is sufficiently clear, but how are the electrons to be divided?

Consider first the ionisation potentials of the half-dozen lightest atoms. The spectra of the hydrogen atom and the helium ion show that in their normal states the single electron occupies a 1s orbit, that is an orbit in which n is 1 and l is 0, the work required to remove it being R and $4R$ respectively. If a second electron is added to the helium ion, it should behave as if attracted by a positive charge of 1 when at large distances, and of 2 when at small distances. Thus the work required to remove the second electron should lie somewhere between R and $4R$ if the electron is bound in a 1s orbit, but be about $R/4$ if it is bound in an orbit in which n is 2. Experiment shows that the ionisation potential of helium is 24·5 volts or $1{\cdot}81R$, so that both electrons occupy 1s orbits and the 'configuration' of the atom may be written $1s^2$.

In this notation orbits are written s, p and d when l assumes the values 0, 1 and 2, small letters being used for the electron in contradistinction to the capitals which describe the atom. In the spectra of the alkalis, which have been explained in terms of a single active electron, the energy of the terms depends primarily on the chief quantum number n and the orbital vector $\mathbf{l}$; accordingly electrons, which have the same value of n and l, will be described as 'equivalent', and a group of equivalent electrons will be said to form a 'shell'.

Only the higher terms of the Li II spectrum are known, but these are similar to the corresponding terms of He I. The normal lithium atom ionises at 5·37 volts or $0{\cdot}40R$, which is rather greater than the $R/4$, which might be expected if the third electron moves in an orbit with an n value of 2 and lying outside the two 1s orbits, but not greater than if it moves in a 2s orbit, which approaches at perihelion nearer to the nucleus than the 1s orbits.

In confirmation the P, D and F states of the lithium atom have very nearly the energies of the stationary states of hydrogen,* and this agrees with theory, for calculations made by Bohr show that the p, d and f orbits lie entirely outside the space in which the first two electrons move. Thus the configuration of lithium is $1s^2 2s$.

The four electron spectra Be I, B II and C III all have a ^1S ground term just as He I has, and one may reasonably assume that the former like the latter arise from an s^2 shell. The ionisation potential of beryllium is greater than that of lithium for the same reasons which make an electron more difficult to remove from helium than hydrogen. The configuration of the normal beryllium atom is thus $1s^2 2s^2$.

The least firmly bound electron of boron is much more easily removed than that of beryllium, so that it can hardly move in a 2s orbit; moreover, the ground term of boron is ^2P, so that the single electron outside the ^1S state of B^+ presumably moves in a p orbit. The normal state of boron is therefore $1s^2 2s^2 2p$, and one supposes that only two electrons can be accommodated in the 2s shell, just as only two seem able to enter the 1s shell. Generalised, this rule states that any s shell is 'complete' when it contains two electrons.†

The periods are strong evidence that the p and d shells are similarly limited in capacity, and the inert gases must surely mark the completion of successive shells. Now the ground terms of all the inert gases are ^1S, so that this may perhaps be the sign of a shell just complete. Certainly the alkaline earths also have ^1S ground terms, which could easily be explained if the first pair of electrons after an inert gas always enter an s shell. A list of spectra having a ^1S ground term has therefore been compiled (Fig. 10·8).

One other observation is suggestive; in the alkalis, when the angular momentum of the electron in its orbit l is 1 in units $h/2\pi$, the orbital moment of the atom is also 1; and conversely when

* For levels of Li I, see Fig. 2·2.

† Much of the above follows Ruark and Urey, *Atoms, molecules and quanta*, 1930. 276.

the atomic state is represented as ^2P, the electron moves in a p orbit. This is true when the core has the configuration of an inert gas; may it not be true whenever the core consists only of complete shells?

In building the first adequate theory of the extra-nuclear structure of the atom, Bohr considered an atom of atomic number Z as built up in Z stages, one electron being added at each stage,

No. of electrons	Spectrum	Shells added beyond last inert gas
2	He I	$1s^2$
4	Be I	$2s^2$
10	Ne I	$2s^2.2p^6$
12	Mg I	$3s^2$
18	A I	$3s^2.3p^6$
20	Ca I	$4s^2$
28	Cu II	$3d^{10}$
30	Zn I	$3d^{10}.4s^2$
36	Kr I	$3d^{10}.4s^2.4p^6$
38	Sr I	$5s^2$
46	Pd I	$4d^{10}$
48	Cd I	$4d^{10}.5s^2$
54	Xe I	$4d^{10}.5s^2.5p^6$
56	Ba I	$6s^2$
70	Lu II	$4f^{14}.6s^2$
78	Au II	$4f^{14}.5d^{10}$
80	Hg I	$4f^{14}.5d^{10}.6s^2$
86	Rn I	$4f^{14}.5d^{10}.6s^2.6p^6$
88	Ra I	$7s^2$

Fig. 10·8. Spectra with a $^1S_0$ ground term; these arise from configurations, in which all the shells are complete.

while the nuclear charge remains unaltered. From each stage a spectrum arises, whose ground term indicates the normal state of the ion, so that when all ionic ground terms are known the atom can be reconstructed; to-day this 'construction principle' remains of value, though the orbits occupied in the ion are not always occupied in the normal atom.

With these tools consider the construction of the atoms period by period. In the first period the ionisation potentials leave no doubt that the configurations of hydrogen and helium are 1s and

$1s^2$ respectively; while in the second the configurations of lithium and beryllium have been shown to be $1s^2\,2s$ and $1s^2\,2s^2$.

In the arc spectrum of boron not many lines are known, but there is no doubt that the system is one of doublets, which resemble the doublets of the alkalis, in all save the appearance of the sharp and diffuse series in absorption. That the ground term is ^2P suggests that the fifth electron occupies a p orbit; and this

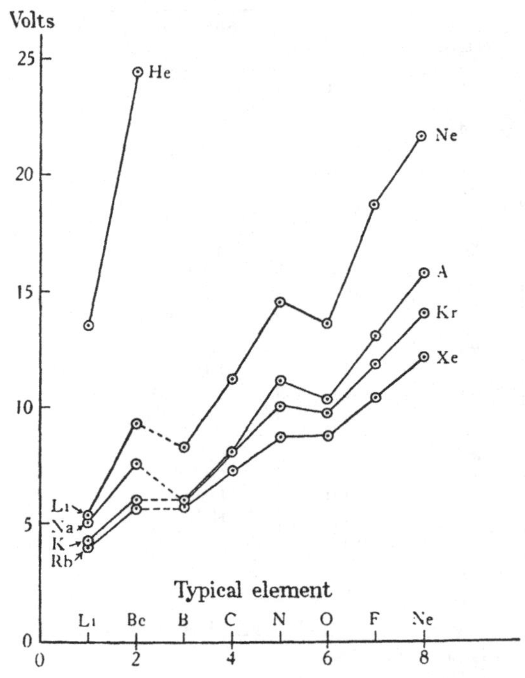

Fig. 10·9. The ionisation potentials of the elements of the short periods.

breaking into a new shell is confirmed by the drop in the ionisation potential from 9·3 volts in beryllium to 8·3 in boron. For the binding of the next five electrons, which produce the elements from carbon to neon, rather more doubtful arguments have to be adduced; the ionisation potential rises from boron to neon, but a slight break between nitrogen and oxygen might suggest two shells instead of one (Fig. 10·9). Negative evidence, however, suggests that all five enter the 2p shell; for must not the orbits all have a chief quantum number of 2? And if so, there is no alternative. In the level diagram of Li I, a 3d ^2D level appears close beside the 3p ^2P level, but no 2d ^2D level appears beside 2p ^2P; similarly, the fine structure of the H_α line has been satisfactorily explained on the assumption that it arises in a transition from states having $n=3$ to states defined by the quantum numbers $n=2$ and $l=0$ or 1; no state, having $n=2$ and $l=2$, has been postulated. Thus it seems necessary to assume that the configuration of carbon is $1s^2 2s^2 2p^2$ and of oxygen $1s^2 2s^2 2p^4$; while if neon marks the completion of the shell, a p shell is limited to six electrons, just as an s shell is limited to two.

No. of electrons outside inert gas	1	2	3	4	5	6	7	8
Spectrum	H	He						
Ground term	^2S	^1S						
Ionisation potential	13·53	24·46						
Spectrum	Li	Be	B	C	N	O	F	Ne
Ground term	^2S	^1S	^2P	^3P	^4S	^3P	^2P	^1S
Ionisation potential	5·37	9·28	8·28	11·212	14·48	13·550	17·34	21·47
Spectrum	Na	Mg	Al	Si	P	S	Cl	A
Ground term	^2S	^1S	^2P	^3P	^4S	^3P	^2P	^1S
Ionisation potential	5·12	7·61	5·96	8·12	11·1	10·31	12·96	15·69
Configuration	s	s^2	s^2 p	$s^2 p^2$	$s^2 p^3$	$s^2 p^4$	$s^2 p^5$	$s^2 p^6$

Fig. 10·10. Ground terms, configurations and ionisation potentials of the short periods.

In passing from neon to sodium the ionisation potential drops from 21·5 to 5·1 volts, so that the valency electron of sodium must enter an orbit much less firmly bound than 2p. Now the third period resembles the second so closely that the configurations must be similar; the ground terms of homologues are identical, and the ionisation potential shows the same changes within the period; thus the configuration of sodium is presumably $1s^2 2s^2 2p^6 3s$ or more briefly (Ne) 3s, where (Ne) signifies the neon core; aluminium is then (Ne) $3s^2$ 3p and argon (Ne) $3s^2 3p^6$ (Fig. 10·10).

In the long periods four ^1S ground terms occur instead of the two of the short periods; appearing in the spectra Ca I, Cu II which is isoelectronic with Ni I, Zn I and Kr I, these divide the fourth period into blocks of two, eight, two and six elements. Bohr enclosed the block of eight in a frame, and as his diagram has been widely used, these eight will be referred to as 'iron frame' elements; the word 'transition' has been widely used to describe these elements, but as it was originally applied to the last three elements, iron, cobalt and nickel only, it sometimes confuses.

Of these blocks the last closely resembles the blocks of six in the short periods; the absorption spectrum of gallium, like that of aluminium, shows the sharp and diffuse series; and the other elements have ground terms identical with those of their short period homologues; the ionisation potentials too show the same characteristic rise with a check between the third and fourth elements, arsenic and selenium. There seems no doubt therefore that these elements arise in the addition of six 4p electrons to the configuration of Zn, in which all the shells are complete.

The first twelve elements of the period present a more difficult problem. With the 3s and 3p shells full in argon, a shell of 3d electrons may reasonably be expected; the first two elements of the fourth period, however, potassium and calcium, have the same ground terms as their homologues, sodium and magnesium, and in other ways resemble them so closely that the nineteenth and twentieth electrons must enter 4s orbits. The next element, scandium, however, has a ^2D ground term, which would suggest that the twenty-first electron enters a 3d orbit even if the spark spectra were not known; but in fact in a system of 19 electrons, the ground term is ^2S only in the first two spectra, K I and Ca II; in the others, Sc III, Ti IV, V V, it is ^2D; so that in this system the 4s orbit is more stable than the 3d only when the nuclear charge is less than 21; when it is 21 or over the 3d orbit is more firmly bound. Of this change Bohr* long ago presented a clear picture; already in lithium the s orbit has been revealed as more firmly bound than the p and d of the same chief quantum number, and this has been explained as due to the greater eccentricity of the s orbit which brings the electron very close to the nucleus at perihelion. The same considerations are valid in the heavier

* Bohr, *Theory of spectra and atomic constitution*, 1922, 95f.

alkalis; in the potassium atom, for example, mathematical analysis shows that the inner loop of the 4s orbit coincides closely with the 3s orbit of hydrogen; and this so alters the dimensions of the outer part of the orbit, that it too deviates from the 4s orbit of hydrogen and coincides instead with a 2s hydrogen orbit, an orbit which is only about a quarter the size. When the nuclear charge is 19, the inner part of the orbit is more stable and the outer part less stable than a circular 3d orbit, but on balance the inner part exerts the greater influence. With increasing nuclear charge however, and the consequent decrease in the difference between the fields of force inside and outside the region occupied by the first 18 electrons, the dimensions of those parts of a 4s orbit which fall outside approach more and more to the dimensions of a 4-quantum orbit calculated on the assumption that the argon core alone shields it from the nucleus. With increasing atomic number therefore a point will be reached where a 3d orbit will be a firmer binding of the nineteenth electron than a 4s orbit. In fact this change occurs when the nuclear charge increases from 20 to 21.

As the ground term of Sc I is ^2D, when the nineteenth electron has occupied a 3d orbit the two next electrons must enter 4s orbits and the configuration of the neutral atom be (A) $4s^2\,3d$. In the succeeding elements the normal state must arise from one of the three configurations $3d^n$, $3d^{n-1}\,4s$ and $3d^{n-2}\,4s^2$, where n is the number of electrons outside the argon core; but no means is available of distinguishing between these alternatives until a method of calculating the terms, which a given configuration produces, has been developed. A ^1S term may however be sought to indicate the completion of the 3d shell; in the arc spectra none occurs before zinc with 12 electrons outside the argon core, but in the spark spectra a ^1S term is basic in both Cu II and Ga II, spectra arising from systems of 10 and 12 electrons respectively. The change from a ^3F ground term in Ni I to a ^1S term in isoelectronic Cu II is so reminiscent of the change from ^2S to ^2D in passing from K I to isoelectronic Sc III that one inclines to ascribe the former to the same cause as the latter, the sinking in the energy scale of the d orbits relative to the s orbits. The ^3F term of Ni I therefore arises presumably from a $3d^9\,4s$ or $3d^8\,4s^2$ configuration, while the ^1S term of Cu II is due to a complete shell of ten d electrons.

To this hypothesis the diagram, which Kossel designed to show the configurations to which ions tend when exhibiting their maximum valency, lends some support; for whereas the elements preceding and succeeding the inert gases tend to assume their configurations, only the elements which succeed nickel, palladium and platinum tend to change; copper is a monovalent metal, but cobalt never an acid radical; zinc readily loses two electrons, but iron never adds two to form an acid similar to sulphuretted hydrogen. And this is just what Bohr's theory suggests, for nickel, palladium and platinum mark the completion of the d shells, and the d orbits grow more stable with increasing nuclear charge. Further, if this hypothesis is true, then the 4s orbits must surely be reoccupied in the two elements succeeding nickel, and in fact copper and zinc have respectively ^2S and ^1S ground terms. Thus we conclude that the configuration of the copper ion is (A) $3d^{10}$ and of neutral zinc (A) $3d^{10}\,4s^2$. In the six remaining elements of the fourth period the electrons enter p orbits, so that the configuration of krypton is (Ne) $3s^2\,3p^6\,3d^{10}\,4s^2\,4p^6$.

Summarising these results (Fig. 10·11) the elements of a long period may be divided into a block of 12 and a block of 6; in the first the outer electrons lie in either the s or d shell, while in the second block the s and d shells are full and the incomplete shell contains only p electrons. But the block of 12 elements may itself be subdivided into three blocks of 2, 8 and 2 elements, the middle block being enclosed by Bohr in a frame; the outer shell contains in the first subdivision only s electrons, in the second a mixture of s and d electrons, and in the third only s electrons. Thus even at this stage it is only the elements in the Bohr frame, whose structure is in serious doubt (Fig. 10·2).

These frame elements possess four properties, which distinguish them from all other elements, save their homologues in the later periods and the rare earths. Their valency is variable; they are active catalysts; and their ions are typically coloured and paramagnetic, since the salts exhibit these properties in

No. of electrons outside inert gas	1	2	3	4	5	6	7	8	9	10	11	12	13	14	15	16	17	18
Spectrum	K	Ca	Sc	Ti	V	Cr	Mn	Fe	Co	Ni	Cu	Zn	Ga	Ge	As	Se	Br	Kr
Ground term	^2S	^1S	^2D	^3F	^4F	^7S	^6S	^5D	^4F	^3F	^2S	^1S	^2P	^3P	^4S	^3P	^2P	^1S
Ionisation potential	4·32	6·09	6·7	6·81	6·76	6·74	7·41	7·83	7·8	7·61	7·68	9·36	5·97	8·09	10·0	9·70	11·80	13·940
Spectrum	Rb	Sr	Y	Zr	Cb	Mo	Ma	Ru	Rh	Pd	Ag	Cd	In	Sn	Sb	Te	I	X
Ground term	^2S	^1S	^2D	^3F	^6D	^7S		^5F	^4F	^1S	^2S	^1S	^2P	^3P	^4S	^3P	^2P	^1S
Ionisation potential	4·16	5·67	6·5	6·92		7·35		7·7	7·7	8·3	7·54	8·96	5·76	7·30	8·73	8·96	10·55	12·078
Probable configuration	s	s^2	$s^2.d$	←——— s and d electrons only ———→							$d^{10}.s$	$d^{10}.s^2$	p	p^2	p^3	p^4	p^5	p^6

Fig. 10·11. Ground terms, configurations and ionisation potentials of the long periods.

dissociated aqueous solution. What is the explanation of these properties in atomic theory? They are catalysts probably because their valency is variable, and their valency varies because the work required to remove an electron is not more than can be supplied in a chemical reaction; in a later chapter this will be related to the almost equal energy content of the s and d orbits. The colour is amenable to a somewhat similar explanation; an ion will be coloured only if it is capable of an electronic transition whose energy is equal to a quantum of visible light, that is of an energy between 1·5 and 3 electron volts or between 35000 and 70000 calories per mole. In an ion, such as Na^+, whose outer shell is complete, the only possible transitions consume so much energy that ultra-violet light is absorbed, and the ion is colourless; in sodium vapour, on the other hand, the atom retains its valency electron, and so absorbs in the yellow as the electron jumps from 3s to 3p. Again, at the end of the Bohr frame copper forms two ions, a cuprous which is colourless and diamagnetic, since the d shell is full, and a cupric which is coloured and paramagnetic, since the d^8 s and d^9 configurations carry roughly equal energy.

The fifth period is like the fourth; all elements outside the Bohr frame have the same ground terms as their homologues, and doubtless therefore similar structures. Within the frames some ground terms are changed, but there is no reason to doubt that the low terms still arise from one of three competing configurations $4d^n$, $4d^{n-1}\,5s$ and $4d^{n-2}\,5s^2$. The most striking difference is the increased stability of the d shell, when it is full; the ground term of Pd I is ^1S, showing that the ten electrons are already settled in the d shell and do not have to wait until an increased nuclear charge brings them in. The monovalent silver ion too, with its system of ten electrons, is so stable that silver exhibits no other valency and forms no coloured ion like copper and gold; in the Ag I spectrum only a single system of doublets is found, just as in the alkalis, whereas in copper and gold there are some extra doublets and some quartet terms; these extra terms are developed from the d^9 configuration of the ion. The ionisation potentials reveal the same contrast, rising sharply from rhodium to palladium and falling again to silver, instead of showing little change as in the fourth period (Fig. 10·12).

In the sixth period electrons may be expected to fill the 4f, 5d and 6s shells; and Bohr's theory clearly predicts that as the nuclear charge increases, there will be one stage when a 5d orbit will be more stable than a 6s, and another when a 4f orbit will be more stable than either; just as there is a nuclear charge which binds the nineteenth electron more firmly in a 3d than in a 4s

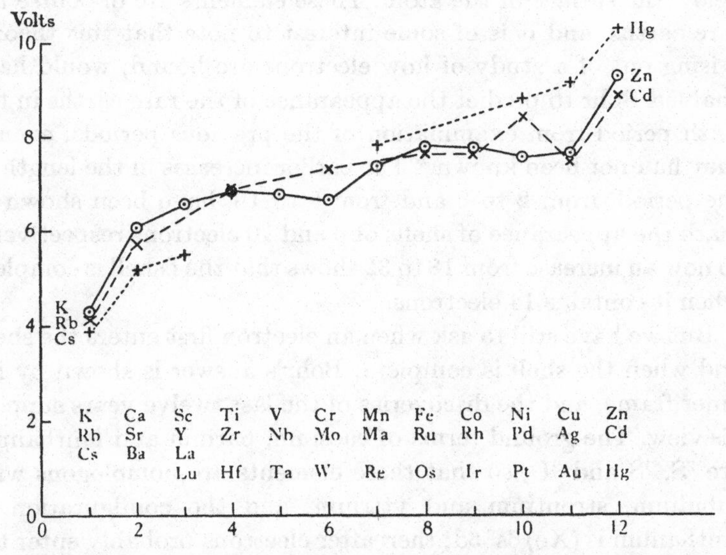

Fig. 10·12. Ionisation potentials of the first twelve elements of the long periods.

orbit, so there must be one nuclear charge which binds an electron more firmly in a 5d orbit than in a 6s, and another which binds it more firmly in a 4f than in a 5d. These predictions Bohr based purely on the shapes of the three orbits, at a time when nothing was known of the ground terms of lanthanum or cerium; they have been confirmed by the work of the last ten years, which has shown that the ground terms of Cs I and Ba II are ^{2}S, and those of La III and Ce IV ^{2}D; so that when there are 55 electrons the last occupies an s orbit when the nuclear charge is 56 or less, and a d orbit when it is 57 or 58. When this electron moves into an f orbit is still not known; many supposed it would be found there in Ce IV, but experiment has proved them wrong.

Early or late, however, there must come a stage in the sixth period when the 4f shell begins to fill, and once this point is reached a number of elements may be expected to follow one another all with nearly the same properties; the similarity should be like the similarity of the iron and palladium frames, but should be more pronounced, for the growing shell now lies farther below the surface of the atom. These elements are of course the rare earths, and it is of some interest to note that this theory, arising out of a study of how electrons are bound, would have enabled Bohr to predict the appearance of the rare earths in the sixth period from examination of the previous periods, even if they had not been known.* The earlier increases in the length of the period, from 2 to 8 and from 8 to 18, have been shown to mark the appearance of shells of 6 and 10 electrons respectively; so now an increase from 18 to 32 shows that the f shell is complete when it contains 14 electrons.

But we have still to ask when an electron first enters the shell, and when the shell is complete; Bohr's answer is shown by his inner frame, and the discoveries of the last twelve years support his view. The ground terms of caesium, barium and lanthanum are ^2S, ^1S and ^2D, so that these elements are homologous with rubidium, strontium and yttrium, and the configuration of lanthanum is (Xe) $6s^2\,5d$; thereafter electrons probably enter the 4f shell until ytterbium is reached. The spectrum of ytterbium itself is not known, but the isoelectronic Lu II certainly has a ^1S ground term, which presumably means that the 16 electrons are arranged as $6s^2\,4f^{14}$. In these 13 rare earths, between cerium and ytterbium, the common valency is always three, and the chemical properties are so similar that they are difficult to separate; while their paramagnetic susceptibilities show that in the trivalent ions all electrons outside the xenon core lie in the f shell. The absorption spectra of the rare earth crystals also deserve attention; the elements of the iron, palladium and platinum frames absorb over bands several hundred angstroms

* Bohr, *Theory of spectra and atomic constitution*, 1922, 110f.

wide, whereas the rare earth bands often cover only 10 A. and at low temperatures much less. If the absorption bands of an ion, whether in a crystal or in solution, are broadened by the electric fields of neighbouring ions, this difference may be taken as some confirmation of theory, for the 4f electrons lying deep within the atom will be little affected by external fields.

Lutecium, which follows ytterbium, is the first of a series which is homologous with the last sixteen elements of the fourth and fifth periods. Lutecium itself is spectroscopically similar to yttrium and lanthanum in that all three have a ^2D ground term; when a seventy-first electron is added to Lu^+, it must therefore enter a 5d orbit, making the configuration of the neutral atom (Xe) $6s^2\,4f^{14}\,5d$. Again, the ground term of Au II with 78 electrons is ^1S, so that it is similar to Cu II and its structure must be (Xe) $4f^{14}\,5d^{10}$; the next two electrons refill the 6s shell, forming ^2S and ^1S ground terms in gold and mercury; while the last six electrons required to make up the full complement of 32 enter p orbits, the ground term of the first element thallium being ^2P and of the last, radon, ^1S. The configuration of radon is thus $(Cu^+)\,4s^2\,4p^6\,4d^{10}\,4f^{14}\,5s^2\,5p^6\,5d^{10}\,6s^2\,6p^6$.

Following the method adopted in the long periods, these facts can be summarised by dividing the rare earth period into five blocks (Fig. 10·13), of which two are Bohr's frames and the other three contain two, two and six elements. Thus the first block contains caesium and barium in which the electrons enter the s shell; the second contains lanthanum and the sequence of eight elements from lutecium to platinum, in which only the s and d shells are incomplete; between lanthanum and lutecium are inserted 13 rare earths, in which the electrons seem to enter f orbits, though in fact the three outer shells may all be incomplete; only in the lutecium ion is the d shell certainly empty.* The remaining eight elements form the fourth and fifth blocks, two electrons refilling the s shell and six filling the p.

In the seventh period only six elements are known, the last being uranium with atomic number 92. In these the neutral atom may be expected to contain seven quantum orbits; in

* The ground terms of Sm and Eu were discovered after this was written.

No. of electrons outside Xe	1	2	3	4	5	6	7	8	9	10	11	12	13	14	15
Spectrum	Cs	Ba	La	Ce	Pr	Nd	Il	Sm	Eu	Gd	Tb	Dy	Ho	Er	Tu
Ground term	^2S	^1S	^2D	^3H				^7F	^8S	^9D					
Ionisation potential	3·87	5·19	5·59	6·54	5·8 R.	6·3 R.		6·5 R.	5·60	6·6 R.	6·7 R.	6·8 R.			
Probable configuration	s	s^2	$s^2.d$	← s, d and f electrons —											

	16	17	18	19	20	21	22	23	24	25	26	27	28	29	30	31	32
	Yb	Lu	Hf	Ta	W	Re	Os	Ir	Pt	Au	Hg	Tl	Pb	Bi	Po	—	Rn
		^2D	^3F		^5D	^6S		^2D	^3D	^2S	^1S	^2P	^3P	^4S	^3P		^1S
	7·1 R.					7·85			8·9	9·2	10·38	6·07	7·38	7·2			10·698
	⟶	$s^2.f^{14}.d$	← f^{14} group complete with varying numbers of s and d electrons →							$f^{14}.d^{10}.s$	$f^{14}.d^{10}.s^2$	p	p^2	p^3	p^4	p^5	p^6

Fig. 10·13. Ground terms, configurations and ionisation potentials of the rare earth period. R. below an ionisation potential means that it was obtained by Rolla and Piccardi from the conductivity of a flame; all others were obtained from analysis of the spectrum.

radium, for example, the two electrons outside the radon configuration move in 7s orbits. The chemical properties of the other four elements are similar to those of the fifth, not the sixth, period; there are, for example, no elements with properties resembling those of the rare earths; the electrons therefore presumably occupy 7s and 6d orbits, leaving the 5f shell still empty.

Thus to-day the structure of most of the elements can be adumbrated from quite simple spectroscopic data. The gaps will be filled in and the argument greatly strengthened in a later chapter, when a method of calculating the terms arising from any group of electrons has been mastered; this method was developed by Hund, from the pioneer work of Russell and Saunders on the displaced terms of the alkaline earths.

The results obtained and those to be obtained later are combined for ease of reference in Fig. 10·14.

6. X-rays

The model of the atom developed in the last section is based on observations of optical spectra; spectra, that is, which arise in the surface layers of the atom. This model the physicist seeks to confirm by asking for evidence of the shells already laid down. This evidence is forthcoming in work on X-rays.

An X-ray emission spectrum consists of lines not unlike optical lines, though the absorption spectrum consists only of edges; both however can be explained as due to a number of energy levels, which when unexcited contain their full complement of electrons. Of these the lowest in the atom is the K shell, though as the energy of an X-ray level is measured as the work required to raise an electron to the periphery of the atom, the K level is also the level of greatest energy. After the K shell come three L levels, L_{I}, L_{II} and L_{III}, in that order, L_{I} requiring more energy to excite it than L_{III}.

To explain even the most complex X-ray spectrum less than two dozen levels are needed, and the number in any shell can be simply related to the number of levels in a doublet system. Thus there is only one K level, and as theory shows that the K shell consists of electrons whose chief quantum number is 1, this

	K	L		M			N				O			P			Q	Ground term
	1s	2s	2p	3s	3p	3d	4s	4p	4d	4f	5s	5p	5d	6s	6p	6d	7s	
H 1	1																	$^2S_{\frac{1}{2}}$
He 2	2																	$^1S_0$
Li 3	2	1																$^2S_{\frac{1}{2}}$
Be 4	2	2																$^1S_0$
B 5	2	2	1															$^2P_{\frac{1}{2}}$
C 6	2	2	2															$^3P_0$
N 7	2	2	3															$^4S_{1\frac{1}{2}}$
O 8	2	2	4															$^3P_2$
F 9	2	2	5															$^2P_{1\frac{1}{2}}$
Ne 10	2	2	6															$^1S_0$
Na 11	10 Ne core			1														$^2S_{\frac{1}{2}}$
Mg 12				2														$^1S_0$
Al 13				2	1													$^2P_{\frac{1}{2}}$
Si 14				2	2													$^3P_0$
P 15				2	3													$^4S_{1\frac{1}{2}}$
S 16				2	4													$^3P_2$
Cl 17				2	5													$^2P_{1\frac{1}{2}}$
A 18	2	2	6	2	6													$^1S_0$
K 19	18 A core						1											$^2S_{\frac{1}{2}}$
Ca 20							2											$^1S_0$
Sc 21						1	2											$^2D_{1\frac{1}{2}}$
Ti 22						2	2											$^3F_2$
V 23						3	2											$^4F_{1\frac{1}{2}}$
Cr 24						5	1											$^7S_3$
Mn 25						5	2											$^6S_{2\frac{1}{2}}$
Fe 26						6	2											$^5D_4$
Co 27						7	2											$^4F_{4\frac{1}{2}}$
Ni 28						8	2											$^3F_4$
Cu^+	2	2	6	2	6	10												$^1S_0$
Cu 29	28 Cu^+ core						1											$^2S_{\frac{1}{2}}$
Zn 30							2											$^1S_0$
Ga 31							2	1										$^2P_{\frac{1}{2}}$
Ge 32							2	2										$^3P_0$
As 33							2	3										$^4S_{1\frac{1}{2}}$
Se 34							2	4										$^3P_2$
Br 35							2	5										$^2P_{1\frac{1}{2}}$
Kr 36	2	2	6	2	6	10	2	6										$^1S_0$
Rb 37	36 Kr core										1							$^2S_{\frac{1}{2}}$
Sr 38											2							$^1S_0$
Y 39									1		2							$^2D_{1\frac{1}{2}}$
Zr 40									2		2							$^3F_2$
Cb 41									4		1							$^6D_{\frac{1}{2}}$
Mo 42									5		1							$^7S_3$
Ma 43									5		2							—
Ru 44									7		1							$^5F_5$
Rh 45									8		1							$^4F_{4\frac{1}{2}}$
Pd 46	2	2	6	2	6	10	2	6	10									$^1S_0$

Fig. 10·14. Electronic configurations of the elements. All the ground terms configuration must often

		K	L		M			N				O			P			Q	Ground term
		1s	2s	2p	3s	3p	3d	4s	4p	4d	4f	5s	5p	5d	6s	6p	6d	7s	term
Ag	47											1							$^{2}S_{\frac{1}{2}}$
Cd	48											2							$^{1}S_{0}$
In	49					46						2	1						$^{2}P_{\frac{1}{2}}$
Sn	50					Pd core						2	2						$^{3}P_{0}$
Sb	51											2	3						$^{4}S_{1\frac{1}{2}}$
Te	52											2	4						$^{3}P_{2}$
I	53											2	5						$^{2}P_{1\frac{1}{2}}$
Xe	54	2	2	6	2	6	10	2	6	10		2	6						$^{1}S_{0}$
Cs	55														1				$^{2}S_{\frac{1}{2}}$
Ba	56					54									2				$^{1}S_{0}$
La	57					Xe core						Xe		1	2				$^{2}D_{1\frac{1}{2}}$
Ce	58										1	core		1	2				$^{3}H_{4}$
Pr	59										2			1	2				
Nd	60										3			1	2				
Il	61										4			1	2				
Sm	62										6				2				$^{7}F_{0}$
Eu	63										7	Xe			2				$^{8}S_{3\frac{1}{2}}$
Gd	64										7	core		1	2				$^{9}D_{2}$
Tb	65					54					8			1	2				
Dy	66					Xe core					9			1	2				
Ho	67										10			1	2				
Er	68										11			1	2				
Tu	69										13				2				
Yb	70										14				2				
Lu^{+++}		2	2	6	2	6	10	2	6	10	14	2	6						
Lu	71													1	2				$^{2}D_{1\frac{1}{2}}$
Hf	72													2	2				$^{3}F_{2}$
Ta	73													3	2				
W	74					68								4	2				$^{5}D_{0}$
Re	75					Lu^{+++} core								5	2				$^{6}S_{2\frac{1}{2}}$
Os	76													6	2				
Ir	77													9					$^{2}D_{2\frac{1}{2}}$?
Pt	78													9	1				$^{3}D_{3}$
Au^{+}		2	2	6	2	6	10	2	6	10	14	2	6	10					$^{1}S_{0}$
Au	79														1				$^{2}S_{\frac{1}{2}}$
Hg	80														2				$^{1}S_{0}$
Tl	81									78					2	1			$^{2}P_{\frac{1}{2}}$
Pb	82									Au^{+} core					2	2			$^{3}P_{0}$
Bi	83														2	3			$^{4}S_{1\frac{1}{2}}$
Po	84														2	4			
—	85														2	5			
Rn	86	2	2	6	2	6	10	2	6	10	14	2	6	10	2	6			$^{1}S_{0}$
—	87																	1	
Ra	88																	2	$^{1}S_{0}$
Ac	89									86							1	2	
Th	90									Rn core							2	2	
Pa	91																3	2	
U	92																4	2	

given have been derived from experiment; in the absence of the ground term the be regarded as doubtful.

corresponds to $1s\,{}^2S_{\frac{1}{2}}$. In all three L levels the chief quantum number is 2, so that they correspond to the optical levels $2s\,{}^2S_{\frac{1}{2}}$, $2p\,{}^2P_{\frac{1}{2},1\frac{1}{2}}$. The optical analogues of the higher X-ray levels, M, N, O, P, are shown in Fig. 10·15.

A line of the K series results from the return of an electron to the K level, but the paucity of lines shows that only certain transitions are permitted. To explain this Wentzel worked out a

k_1	1	2		3		4	
k_2	1	1	2	2	3	3	4
LJ / n	${}^2S_{\frac{1}{2}}$	${}^2P_{\frac{1}{2}}$	${}^2P_{1\frac{1}{2}}$	${}^2D_{1\frac{1}{2}}$	${}^2D_{2\frac{1}{2}}$	${}^2F_{2\frac{1}{2}}$	${}^2F_{3\frac{1}{2}}$
1	K						
2	L I	L II	L III				
3	M I	M II	M III	M IV	M V		
4	N I	N II	N III	N IV	N V	N VI	N VII
5	O I	O II	O III	O IV	O V		
6	P I	P II	P III				

Fig. 10·15. Comparison of X-ray levels, with those of optical doublets.

selection rule based on two quantum numbers k_1 and k_2; to each level certain values of k_1 and k_2 are allotted and an electron can jump from one level to another only if

$$\Delta k_1 = \pm 1 \quad \text{and} \quad \Delta k_2 = 0 \text{ or } \pm 1.$$

These selection rules suggest at once that perhaps k_1 corresponds to the L of optical spectra and k_2 to J; and in fact comparison shows that

$$\left.\begin{array}{l} k_1 = L + \frac{1}{2} \\ k_2 = J + \frac{1}{2} \end{array}\right\}.$$

In the familiar framework, therefore, of a doublet level diagram, the X-ray lines appear in just the same places as the optical lines (Figs. 10·16—17).

This important simplification finds confirmation in measurements of intensity. As one example the intensities of the $K\alpha_1$ and $K\alpha_2$ lines have been measured in a large number of elements and have always been found in the ratio of 2 : 1; now $K\alpha_1$ arises from the transition $L\,\text{III} \rightarrow K$ or ${}^2P_{1\frac{1}{2}} \rightarrow {}^2S_{\frac{1}{2}}$, while $K\alpha_2$ is from

L II → K or $^2P_{\frac{1}{2}} \rightarrow {}^2S_{\frac{1}{2}}$; and the $^2P_{1\frac{1}{2}} \rightarrow {}^2S_{\frac{1}{2}}$ and $^2P_{\frac{1}{2}} \rightarrow {}^2S_{\frac{1}{2}}$ lines of the alkalis have the same 2 : 1 intensity ratio.*

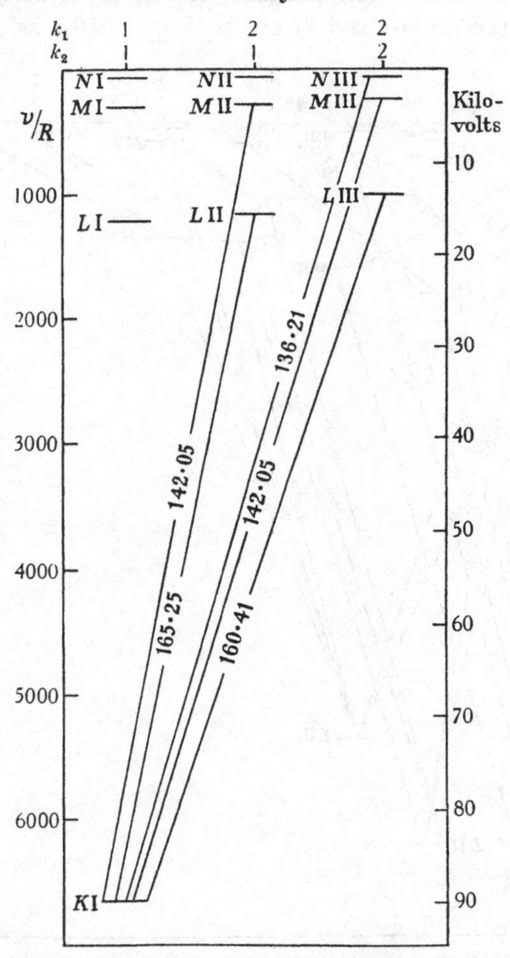

Fig. 10·16. Level diagram showing the K series lines of bismuth; the wavelengths are in X units. ν is used rather inconsistently as an abbreviation for *E/ch*, both here and in the next three pages.

Thus X-ray spectra show that there is one level for each s shell completed, and two for each p, d or f shell. The doublet structure will be taken up in a later chapter with the aid of the vector

* For X-ray intensities see Grebe, *Handbuch der Physik*, 1929, **21** 355.

model; here it is more important to note that, in a graph of $\sqrt{\nu/R}$ against the atomic number, breaks occur when electrons first enter the 3d, 4d and 4f shells (Fig. 10·18); as an electron

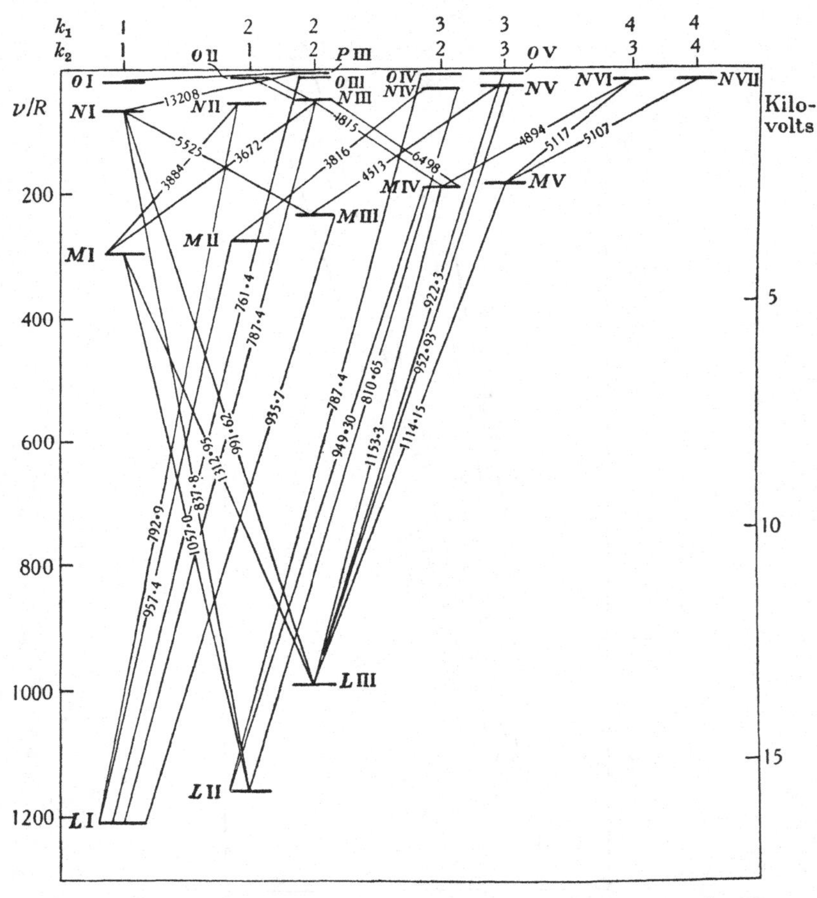

Fig. 10·17. Level diagram showing the L, M, N series lines in the X-ray spectrum of bismuth.

added to one of these shells seems likely to shield the nucleus more than one added to the exterior, the increase in energy between one element and the next should be reduced and the graph should have a smaller slope. In fact regions of decreased slope are observed in the M II, III and L II, III levels between atomic

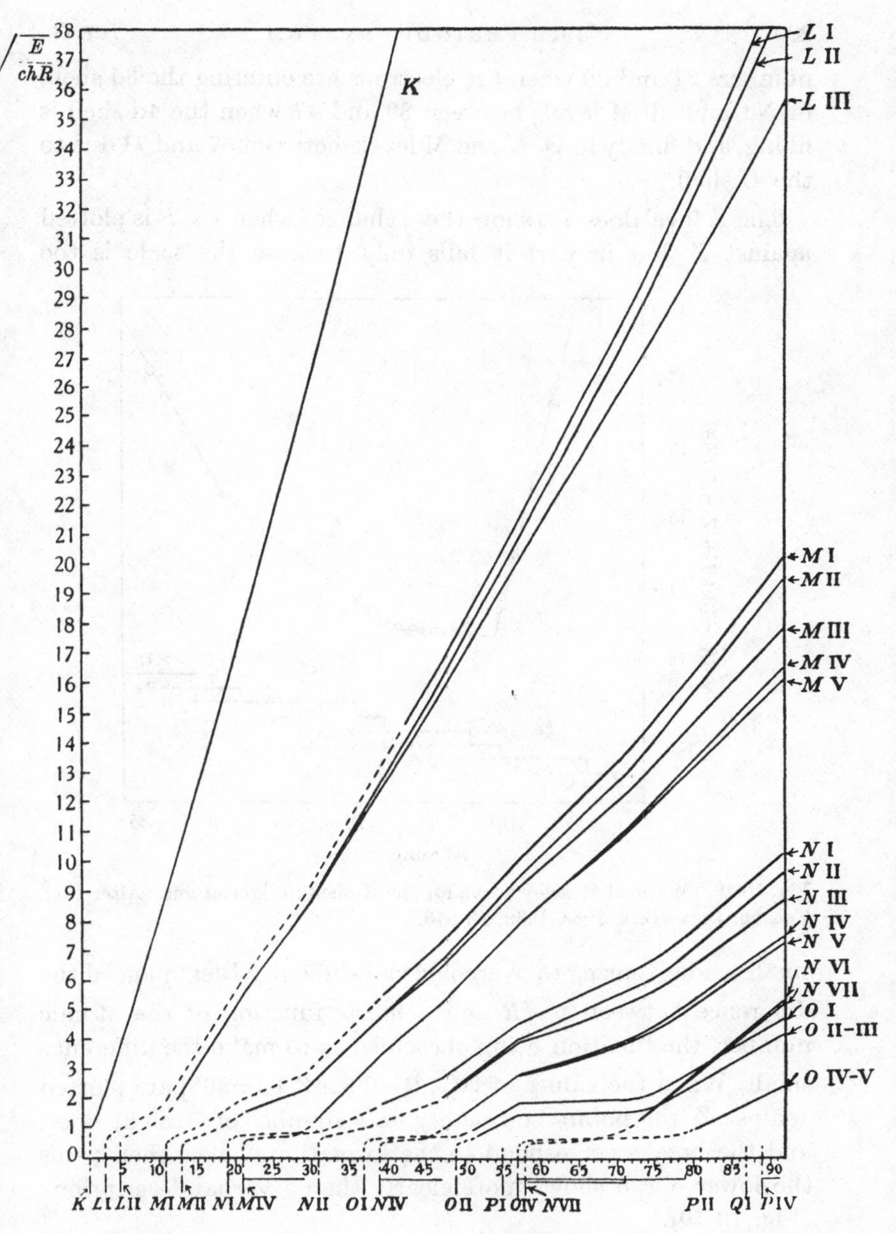

Fig. 10·18. Moseley diagram showing the energies of the X-ray levels plotted against the atomic number. The ordinate is $\sqrt{E/chR}$. After Grotrian, *Graphische Darstellung der Spektren.*

numbers 21 and 29 when the electrons are entering the 3d shell, in N I and all M levels between 39 and 47 when the 4d shell is filling, and finally in O, N and M levels between 57 and 71 due to the 4f shell.

The K level does not show these changes when $\sqrt{\nu/R}$ is plotted against Z, but in part it fails only because the scale is too

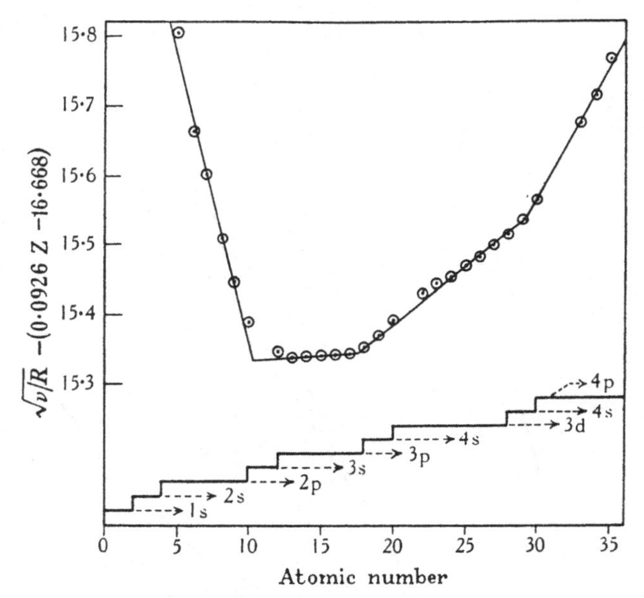

Fig. 10·19. Modified Moseley curve for the K-shell in light atoms. After Idei, *Tohoku. Imp. Univ. Proc.* 1930, **19** 646.

small; endeavouring to overcome this difficulty Idei* plotted the difference between $\sqrt{\nu/R}$ and a linear function of the atomic number, the function being chosen so as to make the difference small. When the values of $\{\sqrt{\nu/R} - 0{\cdot}926Z + 16{\cdot}668\}$ are plotted against Z, the points lie roughly on a number of straight lines, and the breaks correspond to the formation of new shells; this the lower curve shows more clearly than a verbal description† (Fig. 10·19).

* Idei, *Tohoku Univ. Sci. Rep.* 1930, **19** 641.

† For further details see Siegbahn, *Spektroskopie der Röntgenstrahlen*, 1931.

7. Pauli's exclusion principle

The similarities of the columns of the periodic system and the length of its periods have been explained by the simple assumption that there are four types of shell and that the number of electrons which may enter each is limited. Since speculation first began, many have supposed that the inert gases mark the completion of electronic shells, but Pauli* in 1925 was the first to explain why the periods contain 2, 8, 18 or 32 elements.

Name	l	m_l	m_s	Number of orbits	j	m	Number of orbits
s	0	0	$\pm\frac{1}{2}$	2	$\frac{1}{2}$	$\pm\frac{1}{2}$	2
p	1	1 0 -1	$\pm\frac{1}{2}$ $\pm\frac{1}{2}$ $\pm\frac{1}{2}$	6	$\frac{1}{2}$ $1\frac{1}{2}$	$\pm\frac{1}{2}$ $\pm\frac{1}{2}$ $\pm 1\frac{1}{2}$	6
d	2	2 1 0 -1 -2	$\pm\frac{1}{2}$ $\pm\frac{1}{2}$ $\pm\frac{1}{2}$ $\pm\frac{1}{2}$ $\pm\frac{1}{2}$	10	$1\frac{1}{2}$ $2\frac{1}{2}$	$\pm\frac{1}{2}$ $\pm 1\frac{1}{2}$ $\pm\frac{1}{2}$ $\pm 1\frac{1}{2}$ $\pm 2\frac{1}{2}$	10
f	3	3 2 1 0 -1 -2 -3	$\pm\frac{1}{2}$ $\pm\frac{1}{2}$ $\pm\frac{1}{2}$ $\pm\frac{1}{2}$ $\pm\frac{1}{2}$ $\pm\frac{1}{2}$ $\pm\frac{1}{2}$	14	$2\frac{1}{2}$ $3\frac{1}{2}$	$\pm\frac{1}{2}$ $\pm 1\frac{1}{2}$ $\pm 2\frac{1}{2}$ $\pm\frac{1}{2}$ $\pm 1\frac{1}{2}$ $\pm 2\frac{1}{2}$ $\pm 3\frac{1}{2}$	14

Fig. 10·20. Electronic orbits permitted by the exclusion principle.

Pauli postulated that every permitted orbit is determined by four and only four quantum numbers, that orbits having the same quantum numbers are identical, and that no two electrons may at one time occupy the same orbit. If then the quantum numbers are taken to be n, l, m_l and m_s, Pauli's three conditions limit the number of electrons which are permitted for a given value of n, or of n and l; and exactly the same numbers are obtained, if the four quantum numbers chosen are n, l, j and m. This coincidence may surprise at first, but when different quantum numbers are considered as expressions of different types of coupling, it is clear that the number of quantised orbits must be independent

* Pauli, *ZP*, 1925, **31** 765.

of the coupling; indeed this coincidence is an essential condition, which alternative sets of quantum numbers must satisfy.

If n is 1, the only permitted value of l is 0, so that m_l must be 0 and m_s may be $\pm\frac{1}{2}$; while if the other allotment is chosen j must be $\frac{1}{2}$ and m may be $\pm\frac{1}{2}$, so that both allow only two orbits. More generally, if n and l are fixed, there are $(2l+1)$ permitted values of m_l and 2 of m_s, so that there are $2(2l+1)$ orbits in all; and this quantity assumes the values 2, 6, 10, 14 as l increases from 0 to 3, just as experiment requires. The values of m_l and m_s for these states, and the alternative values of j and m, are set out in Fig. 10·20, but the question of which states correspond must be left for the present unsettled.

BIBLIOGRAPHY

The structure of the periodic system was admirably outlined by Bohr in *Theory of spectra and atomic constitution*, 1922, and this is still worth reading. *Atoms, molecules and quanta*, 1930, by Ruark and Urey is a more recent work.

For the chemical properties of the elements and discussions of valency, the soundest works seem to be *The electronic theory of valency*, 1927, and *The covalent link in chemistry*, 1933, both by Sidgwick; *Chemische Bindung als electrostatische Erscheinung*, 1931, by van Arkel and de Boer, also contains much that is valuable.

In *The band spectra of diatomic molecules*, 1932, Jevons provides a lucid empirical approach to the electronic states of molecules; he links these with the Stark levels of an atom and so makes them more easily intelligible to those whose chief interest has been atomic spectra.

In the last few years the quantum mechanics has also been applied to the problems of covalency; Born has reviewed this work in *EEN*, 1931, **10** 387, while some more recent work appears in two papers by Hellmann, *ZP*, 1933, **82** 192, **85** 180. As this book goes to press a paper on 'the quantum theory of valence' by Van Vleck and Sherman appears in *RMP*, 1935, **7** 167.

The standard work on X-rays is *Spektroskopie der Röntgenstrahlen*, 1931, by Siegbahn, while their application to the elucidation of crystal structure is even now being reviewed by Sir William and W. L. Bragg. Vol. **1** of *The Crystalline state* appeared in 1933, two other volumes are promised.

CHAPTER XI
THE DOUBLET LAWS

1. Moseley's law

In 1913 Moseley showed that when the square root of the frequency of any X-ray line is plotted against the atomic number, a straight line results. And what is true for lines is true also for levels, so that the energy E of a given level may be written

$$\sqrt{\frac{E}{chR}}=\frac{1}{n}(Z-s). \qquad \text{......(11·1)}$$

$(Z-s)$ is here referred to as the 'effective nuclear charge', while s, which measures the amount by which the inner electrons seem to reduce the nuclear charge, is called the 'screening constant'. In the K series s is about 2, while in the L series it varies between 10 and 20, values which fit in well enough with the chief quantum numbers assigned to the K and L levels (Fig. 11·1).

Z	Level / Element	K	L I	L III
20	Ca	2·75	—	9·88
55	Cs	3·53	13·93	16·57
82	Pb	1·61	13·61	20·02

Fig. 11·1. The Moseley screening constant in the K and L levels of three elements. For a given level the 'constant' is roughly independent of the atomic number.

Because there is a spectral region ranging from 20 to 500 A. where measurement is very difficult Moseley's law cannot be followed through from the long X-rays, where it still holds fairly well to the optical region. When applied to the outermost electrons it breaks down hopelessly, for interaction between the electrons determines the structure of the spectrum and the variation of nuclear charge is of no account.

Consider, however, a series of spectra arising from ions with the same number of electrons but different nuclear charges, and these difficulties disappear; the interaction between electrons will remain roughly the same, while the effective nuclear charge changes. And in fact Bowen and Millikan* pointed out in 1924 that if $\sqrt{E_l/chR}$ be plotted against Z in an isoelectronic sequence, then a straight line results.

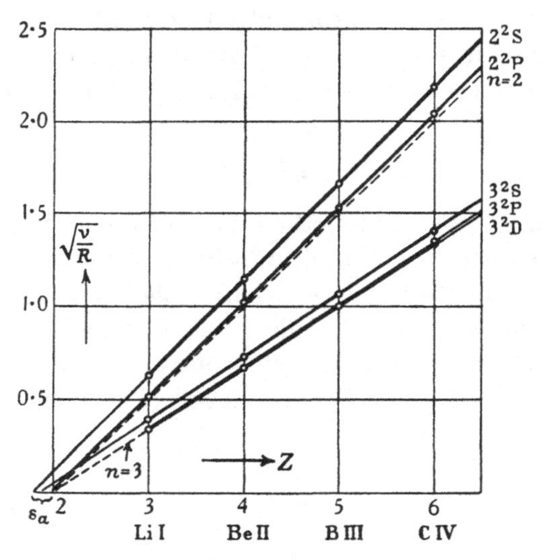

Fig. 11·2. Moseley diagram for a system of three electrons. The dotted lines show the positions of hydrogen terms with increasing nuclear charge. After Grotrian, *Graphische Darstellung der Spektren.*

Fig. 11·2 shows the result for the sequence Li I, Be II, B III and C IV, each spectrum arising from a system of three electrons. If in this system the two K electrons lay so close to the nucleus that they reduced the nuclear charge by two, but had no other influence on the optical electron, the spectra would be identical with those of the sequence H I, He II and Li III; the dotted line would then show the terms having $n = 2$, and it is worthy of notice that though the 2 ^{2}S and 2 ^{2}P terms are slightly displaced, they are closely parallel to this line. The 3 ^{2}D term is so like the

* Bowen and Millikan, *PR*, 1924, **24** 209.

hydrogen line having $n=3$ that the two cannot be drawn separate.

When more electrons are present larger deviations may be expected, and in fact these Moseley diagrams show very clearly how terms change places in the energy sequence with a change of nuclear charge. The Moseley diagram for a system of 19 electrons, Fig. 11·3, shows, for example, how the line of the 3d ^{2}D term cuts

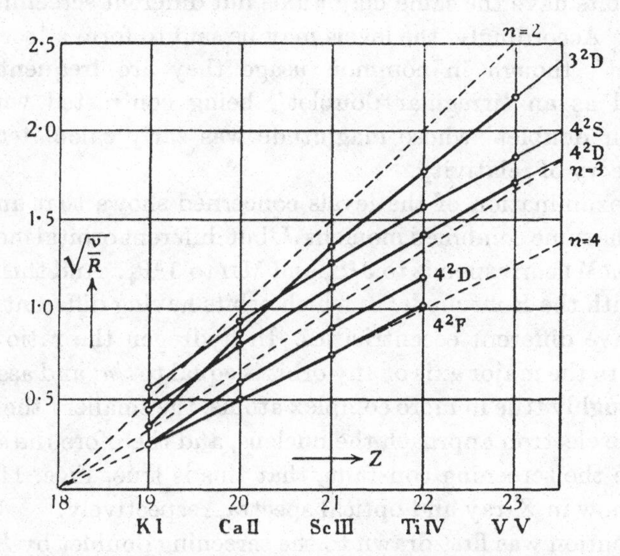

Fig. 11·3. Moseley diagram for a system of 19 electrons. Note how the 4s ^{2}S level crosses the 3d ^{2}D level. After Grotrian, *Graphische Darstellung der Spektren.*

the 4s ^{2}S line when the nuclear charge changes from 20 to 21, and this is not unimportant, for the 3d electron sinks relative to the 4s electron throughout the iron row, whenever the nuclear charge is increased.

These Moseley diagrams have been widely used in the elucidation of spectra; thus Fowler and Selwyn,* in 1928, having the spectra of O III and N II before them, used this method in their analysis of C I. Some of these diagrams have been published by Grotrian.

* Fowler, A. and Selwyn, *PRS*, 1928, **118** 46.

2. Screening doublets

In the Moseley X-ray diagrams, shown in Fig. 10·18, certain levels such as $L_I L_{II}$ and $M_I M_{II}$ run parallel. Since the equation of a line is

$$\sqrt{\frac{E}{chR}} = \frac{1}{n}(Z-s), \qquad \text{......(11·1)}$$

the slope is determined by n and the position by s; that is, the two orbits have the same major axis but different screening constants. Accordingly, the levels may be said to form a 'screening doublet', though in common usage they are frequently described as an 'irregular doublet', being contrasted with the 'regular doublet' whose magnitude was early calculated from the theory of relativity.

An examination of the levels concerned shows that any pair have the same combined moment J, but different orbital moments L; thus M_I corresponds to $3\,^2S_{\frac{1}{2}}$ and M_{II} to $3\,^2P_{\frac{1}{2}}$. And this agrees well with the Bohr model in which orbits having different values of l have different eccentricities. In hydrogen the ratio of the minor to the major axis of any orbit is equal to l/n; and assuming this roughly true in more complex atoms, the smaller l the nearer will the electron approach the nucleus, and therefore the smaller will be the screening constant; that this is true, Figs. 11·1 and 11·2 show in X-ray and optical spectra, respectively.

Attention was first drawn to the screening doublet by Hertz,* who had observed parallel lines in a level scheme. Similar parallel lines cannot be observed when $\sqrt{\nu/R}$ is plotted against the atomic number for a number of lines, because the selection rules do not allow both levels of a screening doublet to combine with a third level.

On the other hand, the two levels are permitted to combine together; among the low X-ray levels, combinations involving no change in n have not indeed been observed, but probably this is only because they are of low intensity, for lines have been identified as transitions between various N levels, between N_{IV} and N_{VI} in gold for example,† while optical levels, which give parallel Moseley lines, often combine.

* Hertz, *ZP*, 1920, **3** 19. † Idei, *N*, 1929, **123** 643.

The combination of these optical levels produces lines whose frequency is a linear function of the atomic number; for in a set of isoelectronic spectra a pair of Moseley lines may be written

$$\sqrt{\frac{E_1}{chR}}=\frac{1}{n_1}(Z-s_1)$$

and
$$\sqrt{\frac{E_2}{chR}}=\frac{1}{n_2}(Z-s_2),$$

and if the lines are parallel $n_1=n_2$. Accordingly, the frequency of the resulting line is

$$\frac{\nu}{R}=\frac{E_2-E_1}{chR}=\frac{2(s_1-s_2)Z+(s_2^2-s_1^2)}{n_1^2}. \quad ...(11{\cdot}2)$$

Millikan and Bowen have used this relation in finding the $2\,^2P_{\frac{1}{2}}\to 1\,^2S_{\frac{1}{2}}$ line of many 'stripped' atoms, that is, of ions which

Spectrum	Wave-length	Wave-number	Difference
Na I	5895·930	16,956	
			18,713
Mg II	2802·70	35,669	
			17,993
Al III	1862·90	53,662	
			17,593
Si IV	1402·9	71,255	
			17,358
P V	1128·039	88,613	
			17,206
S VI	944·590	105,819	
			17,123
Cl VII	813·00	122,942	

Fig. 11·4. 3p $^2P_{\frac{1}{2}}\to$3s $^2S_{\frac{1}{2}}$ screening doublet in a system of 11 electrons. The difference does not change much, and its small change is regular.

have only one electron outside an inert gas shell and are therefore isoelectronic with the alkalis. The frequency of this line in a number of isoelectronic spectra is a linear function of the atomic number, and accordingly when the wave-length of the line in Na I and Mg II is known, its wave-length in Al III, Si IV, P V, S VI and Cl VII may be calculated. This law does not hold with the high accuracy of spectroscopic work, as Fig. 11·4 shows, but the approximate position is often sufficient to identify the line, for the $2\,^2P_{\frac{1}{2}}\to 1\,^2S_{\frac{1}{2}}$ line must be one of the brightest in the spectrum and the doublet separation $2\,^2P_{1\frac{1}{2}}-2\,^2P_{\frac{1}{2}}$, which can be calculated from the spin doublet law, is a valuable aid. In fact, so successful has the method been that this doublet has been measured from the alkalis as far as O VI, Cl VII, Mn VII, Zr IV and Pr V in successive periods.*

Spectrum	Wave-length	Wave-number	Difference
Mg I	2852·11	35,051	
			24,779
Al II	1670·81	59,830	
			22,994
Si III	1206·9	82,824	
			22,320
P IV	950·66	105,144	
			21,939
S V	786·51	127,083	

Fig. 11·5. 3s 3p $^1P_1 \rightarrow 3s^2\ ^1P_0$ screening doublet in a system of 12 electrons.

Spectrum	Wave-length	Wave-number	Difference
Mg I	5172·70	19,327	
			28,700
Al II	2081·5	48,027	
			52,412
Si III	995·20	100,439	
			58,455
P IV	629	158,894	
			69,156
S V	438·19	228,050	

Fig. 11·6. The screening doublet law is no longer valid for the 3s 4s $^3S_1 \rightarrow$ 3s 3p $^3P_1$ line in a system of 12 electrons.

In optical spectra, levels which give parallel Moseley lines do not necessarily have the same value of J, for the square root of the frequency of the $2\,^1P_1 \rightarrow 1\,^1S_0$ line is a linear function of the atomic number, as Fig. 11·7 shows; but the condition that the chief quantum shall not change appears to be still valid, for the square root of the wave-number of the 3s 4s $^3S_1 \rightarrow$ 3s 3p $^3P_1$ line of magnesium is not a linear function of the atomic number (Fig. 11·6).

3. Spin doublets

The pairs of adjacent Moseley lines in the X-ray level diagrams (Fig. 10·18), which are not parallel, diverge rapidly with in-

* Gibbs and White, *PR*, 1929, **33** 159.

creasing atomic number. These pairs have been known as relativity or regular doublets, because Sommerfeld explained them as due to a relativistic change in the mass of the electron, at a time when the 'irregular' or screening doublets still defied explana-

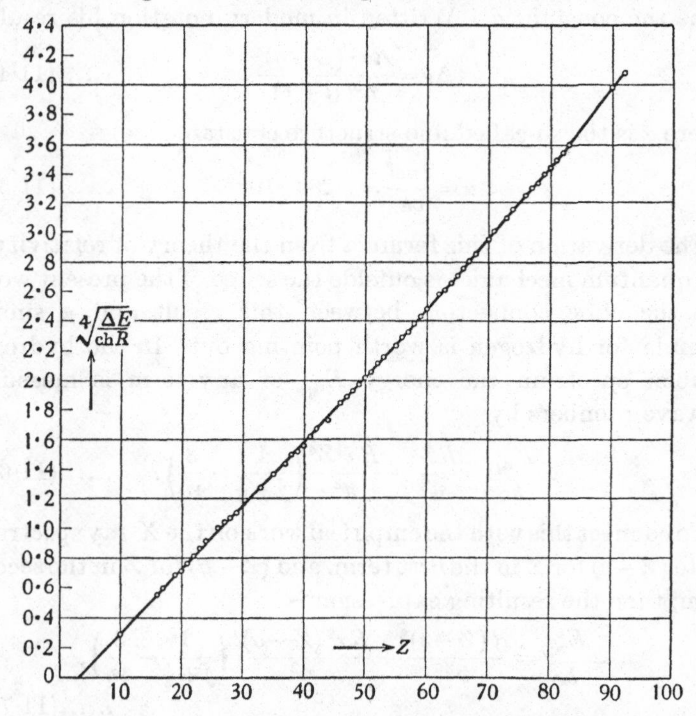

Fig. 11·7. The spin doublet law applied to the LII LIII X-ray levels; some of the level values on which this figure is based appear in Fig. 11·8. After Grotrian, *Graphische Darstellung der Spektren.*

tion; in the model, however, they are to be ascribed to the magnetic interaction of the spin and orbital vectors.

The terms composing a spin doublet* have the same value of the orbital moment L, but values of J differing by unity. When $\sqrt[4]{\Delta\nu/R}$ is plotted against Z a straight line results, showing that the interval $\Delta\nu$ may be written

$$\Delta\nu = d' R (Z-\sigma)^4, \qquad \text{.....(11·3)}$$

* In this nomenclature I follow Pauling and Goudsmit, *Structure of line spectra*, 1930, 181.

where d' determines the slope of the line in Fig. 11·7 and σ is a new screening constant.

Sommerfeld, using the relativistic change of mass with velocity, was able to develop this formula mathematically and to determine the constant d'. Written in modern notation his result is

$$\Delta\nu = \frac{R\alpha^2 (Z-\sigma)^4}{n^3 l(l+1)}, \qquad \text{......(11·4)}$$

where α is the so-called fine-structure constant,

$$\alpha = \frac{2\pi e^2}{ch} = 7{\cdot}284 \, . \, 10^{-3}. \qquad \text{......(11·5)}$$

The derivation of this formula from the theory of relativity or the quantum mechanics is outside the scope of the present work; but the close connection between this result and a similar formula for hydrogen is worth pointing out. In the hydrogen doublet spectrum the energy $E_{n_{lj}}$ of any term is measured in wave-numbers by

$$-\frac{E_{n_{lj}}}{hc} = \frac{RZ^2}{n^2} + \frac{R\alpha^2 Z^4}{n^3}\left(\frac{1}{j+\frac{1}{2}} - \frac{3}{4n}\right). \qquad \text{......(11·6)}$$

To connect this with the empirical work on the X-ray spectrum, write $(Z-s)$ for Z in the first term, and $(Z-\sigma)$ for Z in the second term, then the resulting expression is

$$-\frac{E_{n_{lj}}}{hc} = \frac{R(Z-s)^2}{n^2} + \frac{R\alpha^2 (Z-\sigma)^4}{n^3}\left(\frac{1}{j+\frac{1}{2}} - \frac{3}{4n}\right), \qquad \text{......(11·7)}$$

and this is very similar to that obtained by adding to the energy of a term obeying Moseley's law the energy of a relativity doublet.

Moreover, this expression, when used to calculate the separation of two doublet levels having the same values of n, l, s and σ, but different values of J, yields the Sommerfeld formula (11·4) again; for in a doublet spectrum the permitted values of J are $(l+\frac{1}{2})$ and $(l-\frac{1}{2})$.*

The Sommerfeld formula contains only one quantity not determined, namely the screening constant σ, and this may

* Goudsmit and Uhlenbeck, *Physica*, 1926, **6** 273.

therefore be calculated from the splitting of any doublet. Calculations based on the simple formula of equation (11·4) do not make σ constant (Fig. 11·8); but theory dictates certain correction terms, whose discussion is outside the scope of the present work,* and if these are included the constancy of σ is very striking. Indeed, the screening constant of the L II L III varies only between 3·42 and 3·56 in a series of fifty elements; and recently Zahrahnicek has been able to express even this small variation as a function of the atomic number.†

Element	Z	E/chR		$\sqrt[4]{\frac{\Delta E}{chR}}$	σ	
		L II	L III		By equation (11·4)	With Sommerfeld's correction term
Ca	20	25·9	25·6	0·74	2·7	2·7
Zn	30	76·81	75·14	1·14	3·29	3·5
Zr	40	170·0	163·8	1·58	3·03	3·45
Sn	50	306·2	289·5	2·02	2·67	3·50
Nd	60	495·5	457·8	2·478	1·95	3·51
Yb	70	735·4	659·2	2·955	0·76	3·5
Hg	80	1048·6	906·1	3·455	−0·93	3·44
Th	90	1451·5	1200·6	3·982	−3·30	3·43

Fig. 11·8. Sommerfeld's doublet formula applied to the L II L III X-ray levels.

The great success of the formula in dealing with numerical values fosters the belief that the Sommerfeld derivation is correct, but after the introduction of the spinning electron precisely the same formula was developed by the methods of the quantum mechanics,‡ the mechanism implied being the interaction of the spin and orbital moments. The difference, however, is perhaps more apparent than real, for Dirac has derived the spin of the electron from relativistic equations.

When applied to an isoelectronic sequence of optical spectra the simple Sommerfeld formula of equation (11·4) is less satisfactory; if it is tested by calculating σ as before, σ is seen to decrease as the atomic number increases. But among ions iso-

* Sommerfeld, *Atomic structure and spectral lines*, 1923, 500.

† Zahrahnicek, *ZP*, 1930, **60** 712.

‡ Darwin, C. G., *PRS*, 1927, **115** 1.

electronic with the alkalis the trend can be easily traced, as Fig. 11·9 shows, and accordingly Millikan and Bowen* used the formula to estimate the $2\,{}^2P_{1\frac{1}{2}}-2\,{}^2P_{\frac{1}{2}}$ interval in the higher spark spectra of the short periods.

Spectrum	$2p\ \Delta^2P_{\frac{1}{2},1\frac{1}{2}}$	σ	Spectrum	$3p\ \Delta^2P_{\frac{1}{2},1\frac{1}{2}}$	σ
	cm.$^{-1}$			cm.$^{-1}$	
Li I	0·338	2·018	Na I	17·18	7·447
Be II	6·61	1·935	Mg II	91·55	6·601
B III	34·1	1·888	Al III	238·00	6·045
C IV	107·4	1·855	Si IV	461·84	5·909
N V	259·1	1·835	P V	794·82	5·732
			S VI	1267·10	5·587
			Cl VII	1889·5	5·492

Fig. 11·9. The spin doublet law applied to alkali-like spectra.

Spectrum	$2s.2p\ \Delta^3P_{0,2}$	σ	Spectrum	$3s.3p\ \Delta^3P_{0,2}$	σ
	cm.$^{-1}$			cm.$^{-1}$	
Be I	3·02	2·303	Mg I	60·81	7·126
B II	22·8	2·186	Al II	187·3	6·543
C III	79·7	2·152	Si III	373	6·330
N IV	204·1	2·133	P IV	696·5	6·034
O V	441	2·100	S V	1120	5·904
			Cl VI	1702	5·790

Fig. 11·10. The spin doublet law applied to alkaline earth-like spectra.

Further, the Sommerfeld formula will predict the overall triplet interval of the alkaline earths with much the same accuracy (Fig. 11·10). But though σ is regular enough within a single isoelectronic sequence, it varies from sequence to sequence in a wholly irregular manner. Moreover, though the spin doublet formula has been used on the short periods, Gibbs and White‡ have found that in the long periods the Landé doublet formula gives more satisfactory results.

4. Landé's doublet formula

Sommerfeld developed his doublet formula on the assumption that an electron moves in a uniform field equal to that of a constant point charge $(Z-\sigma)e$. This is a reasonable assumption

* Millikan and Bowen, *PR*, 1926, **27** 144.
† Gibbs and White, *PR*, 1929, **33** 157.

for an X-ray orbit deep within the atom, but it hardly describes the forces acting on an electron in an outer orbit. Accordingly, Landé divided the orbit into two parts, an inner one in which the electron was subject to the attraction of a charge $Z_i e$,

Spectrum	Configuration	n^*	Interval cm.$^{-1}$	Z	Z_i	$Z - Z_i$
			$\Delta^2P_{\frac{1}{2},1\frac{1}{2}}$			
Li	p	1·96	0·34	3	0·9	2·1
Na		2·12	17·18	11	7·5	3·5
K		2·23	57·71	19	14·8	4·2
Rb		2·28	237·6	37	31·1	5·9
Cs		2·33	554·1	55	49·1	5·9
B	s^2.p	1·32	15	5	3·5	1·5
Al		1·51	112·01	13	11·5	1·5
Ga		1·51	826·0	31	31·3	−0·3
In		1·53	2213	49	52·2	−3·2
Tl		1·49	7793	81	94·1	−13·1
Cu	d^{10}.p	1·86	248·4	29	23·4	5·6
Ag		1·87	920·6	47	45·5	1·5
Au		1·72	3816	79	81·7	−2·7
			$\Delta^3P_{0,2}$			
Mg	sp	1·66	60·76	12	9·8	2·2
Ca		1·79	158·1	20	17·7	2·3
Sr		1·86	581	38	35·8	2·2
Ba		1·92	1249	56	55·1	0·9
Zn	d^{10}.sp	1·59	578·7	30	28·3	1·7
Cd		1·63	1713	48	50·5	−2·5
Hg		1·53	6398	80	88·7	−8·7
			$\Delta^5P_{1,3}$			
O	p^4 or	2·17	6·1	8	4·6	3·4
S	p^3 (^{4}S) p	2·33	29·16	16	11·3	4·7
Se		2·38	148·5	34	26·2	7·8
			$\Delta^2P_{2,0}$			
F	p^5	0·85	407·0	9	9·3	−0·3
Cl		1·02	881	17	17·9	−0·9
Br		1·07	3685	35	39·4	−4·4

Fig. 11·11. Landé's doublet formula applied to the lowest P term in various columns of the periodic system.

an outer where the attracting charge was $Z_a e$. Z_i should then be slightly less than the atomic number Z, and Z_a will be unity for a normal atom, 2 for a singly ionised atom, and will assume higher integral values if still more electrons are removed.

Working on these lines Landé obtained for the separation of an alkali doublet the value

$$\Delta\nu = \frac{R\alpha^2 Z_i^2 Z_a^2}{n^{\star 3} l(l+1)}. \qquad \ldots\ldots(11{\cdot}8)$$

This clearly becomes identical with the Sommerfeld formula, if Z_a is made equal to Z_i and if for the effective quantum number

Spectrum	Configuration	$n^\star$	Interval cm.$^{-1}$	Z	Z_i	$Z - Z_i$
			$\Delta\ ^2P_{\frac{1}{2}, 1\frac{1}{2}}$			
Na I	3p	2·12	17·18	11	7·5	3·5
Mg II		2·26	91·55	12	9·5	2·5
Al III		2·37	238·0	13	11·0	2·0
Si IV		2·45	461·8	14	12·1	1·9
P V		2·5	794·8	15	13·1	1·9
S VI		2·55	1267	16	14·2	1·8
			$\Delta\ ^3P_{0,2}$			
Mg I	3s.3p	1·66	60·8	12	9·8	2·2
Al II		1·96	187·3	13	11·0	2·0
Si III		2·13	373	14	11·7	2·3
P IV		2·25	696·3	15	13·0	2·0
S V		2·34	1120	16	14·1	1·9

Fig. 11·12. Landé's doublet formula applied to the lowest P term in systems of 11 and 12 electrons.

$n^\star$ is substituted the true quantum number n. In this equation $n^\star$ is determined as usual by the equation

$$E_l/ch = \frac{RZ_a^2}{n^{\star 2}}, \qquad \ldots\ldots(11{\cdot}9)$$

where E_l is the energy of the doublet terms measured from the sequence limit. The formula is interesting chiefly because it collates three empirical rules, previously unconnected.

Long ago Kayser and Runge pointed out that in a series of homologous elements, such as lithium, sodium and potassium, a given interval increases rather more rapidly than the square of the atomic number. According to Landé's formula this interval should vary as $Z_i^2/n^{\star 3}$, or roughly as Z_i^2, since $n^\star$ will change but little (Fig. 11·11).

Secondly, in a sequence of isoelectronic spectra, neither Z_i nor $n^\star$ will vary greatly, so that the interval will be roughly proportional to Z_a^2. The formula thus extends the well-known rule that isoelectronic arc and spark spectra are similar but drawn on scales whose ratio is 1 : 4 (Fig. 11·12).

Thirdly, inside a single term sequence the interval decreases inversely as n^{*3} (Fig. 11·13).

Spectrum	Con-figuration	n^*	Interval cm.$^{-1}$	Z	Z_i
			$\Delta^2P_{\frac{1}{2},1\frac{1}{2}}$		
Na I	3p	2·12	17·18	11	7·5
	4p	3·13	5·49		7·6
	5p	4·14	2·49		7·7
Cs I	6p	2·33	554·1	55	49·1
	7p	3·39	181·1		49·2
	8p	4·40	80·6		48·6
	9p	5·40	45·1		49·4
			$\Delta^2D_{1\frac{1}{2},2\frac{1}{2}}$		
Cs I	5d	2·55	97·59	55	40·8
	6d	3·53	42·8		44·1
	7d	4·53	20·9		44·7
	8d	5·53	11·6		45·0
			$\Delta^2P_{\frac{1}{2},1\frac{1}{2}}$		
Al I	$3s^2.3p$	1·51	112·01	13	11·5
	$3s^2.4p$	2·67	15·22		10·0
	$3s^2.5p$	3·70	5·95		10·2
	$3s^2.6p$	4·71	2·82		10·1
			$\Delta^3P_{0,2}$		
Zn I	4s.4p	1·59	578·7	30	28·3
	4s.5p	2·75	82·92		24·3
	4s.6p	3·78	30·91		23·9
	4s.7p	4·79	15·05		23·8

Fig. 11·13. Landé's doublet formula applied to various series.

But the formula does more than summarise these three rules, for it may be used to determine Z_i from the relation

$$Z_i = \frac{1}{\alpha Z_a}\sqrt{\frac{n^{*3}\, l\,(l+1)\,\Delta\nu}{R}}. \qquad \ldots\ldots(11\cdot10)$$

And in general the values of Z_i thus obtained are reasonable as well as consistent. Thus for P terms we might expect that $(Z-Z_i)$ would lie between 2 and 10, the active electron being screened by the K shell and perhaps by part of the L shell as well. And in fact in a large number of spectra, Z_i does lie between these limits. Again, we might expect that in the D terms $(Z-Z_i)$

would be rather greater than 10, and in fact the values obtained, though covering rather a wide range, are of this order.

Besides these successes, the Landé formula is able to predict the intervals of the alkaline earth triplets, just as the Sommerfeld relativity formula does; figures illustrating the point are given here, but the explanation must wait until the manner in which electrons combine has been described.

Gibbs and White have used the formula to calculate the ^2P intervals of the higher spark spectra of the long periods.

BIBLIOGRAPHY

Grotrian, *Graphische Darstellung der Spektren von Atomen mit ein, zwei und drei Valenzelektronen*, 1928, **1** contains a discussion, and vol. **2** contains several diagrams. The treatment by Pauling and Goudsmit in *The structure of line spectra*, 1930, is more mathematical.

APPENDIX I

KEY TO REFERENCES

The periodicals, in which most papers appear, I have cited by the capitals introduced by Gibbs, the less common by the usual abbreviations. A key to the former is given here; a key to the latter may be found in *Science Abstracts*.

AJ	Astrophysical Journal.
AP	Annalen der Physik.
BSJ	Bureau of Standards, Journal of Research.
EEN	Ergebnisse der Exacten Naturwissenschaften.
JOSA	Journal of the Optical Society of America.
N	Nature.
Nw	Naturwissenschaften.
PM	Philosophical Magazine.
PR	Physical Review.
PRS	Proceedings of the Royal Society, London, series A.
PZ	Physikalische Zeitschrift.
ZP	Zeitschrift für Physik.

An author's initials are not given in the text unless two authors of the same name occur; but all authors are given their initials in the index.

APPENDIX II

SOME PHYSICAL CONSTANTS

The values of the physical constants used throughout this book are those recommended by Birge* in 1929, but with the divergence between the deflection and spectroscopic values of e/m_e resolved. While the book was in the press, however, the second of two further papers by Birge† appeared; this suggests a change in the accepted values of the fundamental constants e and h by no less than 0·7 and 1·2 per cent. respectively, with consequent changes in all the derived constants. The change is due to a new determination of the viscosity of air, compelling a recalculation of Millikan's work on the charge of an electron. At so late a stage it seemed unwise to change the constants throughout the book; the experimental results of too many different workers would have to be recalculated; so the old values have been left standing and the new values added in brackets.

Spectra

Velocity of light

$c = 2{\cdot}9979_6 \,.\, 10^{10}$ cm. sec.$^{-1}$

Rydberg constant for hydrogen

$R_H = 109677{\cdot}759 \pm 0{\cdot}05$ cm.$^{-1}$

Rydberg constant for helium

$R_{He} = 109722{\cdot}403 \pm 0{\cdot}05$ cm.$^{-1}$

Rydberg constant for infinite mass

$R_\infty = 109737{\cdot}42 \pm 0{\cdot}06$ cm.$^{-1}$

Sommerfeld's fine structure constant

$\alpha = \dfrac{2\pi e^2}{hc} = 7{\cdot}284 \,.\, 10^{-3}$ (7·296)

$\alpha^2 = 5{\cdot}305 \,.\, 10^{-5}$ (5·323)

$1/\alpha = 137{\cdot}4$ (137·06)

Spectroscopic doublet constant

$R_\infty \alpha^2 = 5{\cdot}821$ cm.$^{-1}$ (5·841)

Zeeman displacement

$o_m/H = \dfrac{e}{4\pi m_e c^2} = 4{\cdot}665 \,.\, 10^{-5}$ cm.$^{-1}$ gauss^{-1}

Energy

Mechanical equivalent of heat

$J = 4{\cdot}185_2$ joule cal.$^{-1}$

1 electron volt excites a line of wave-number

8106 cm.$^{-1}$ (8057)

* Birge, *Rev. Mod. Phys.* 1929, **1**, 59.

† Birge, *Science*, 1934, **79**, 438; *PR*, 1935, **48**, 918.

1 electron volt excites a line of wave-length

12,336 A. (12,412)

Energy of 1 electron volt

$1{\cdot}5911 . 10^{-12}$ erg (1·6012)

1 electron volt molecule^{-1}

23,055 cal. mol.$^{-1}$ (23,198)

Planck's constant

$h = 6{\cdot}54_7 . 10^{-27}$ erg sec. (6·6286)

Boltzmann's constant

$k = 1{\cdot}372 . 10^{-16}$ erg °C.$^{-1}$

Atoms and electrons

Avogadro's number

$N_0 = 6{\cdot}064 . 10^{23}$ mol.$^{-1}$

Distance between nucleus and electron in normal hydrogen orbit

$a_0 = 0{\cdot}5284 . 10^{-8}$ cm. (0·5289)

Bohr unit of angular momentum

$h/2\pi = 1{\cdot}042 . 10^{-27}$ erg sec. (1·055)

Bohr magneton

$\mu_B = \dfrac{eh}{4\pi m_e c} = 0{\cdot}9156 . 10^{-20}$ erg gauss^{-1} (0·9271)

Magnetic moment per mol for one Bohr magneton per molecule

$\mu_B N_0 = 5552 \pm 10$ erg gauss^{-1} mol.$^{-1}$ (5622)

Charge: mass ratio of electron:

$e/m_e c = 1{\cdot}757_5 . 10^7$ e.m.u. gm.$^{-1}$

Electronic charge

$e = 4{\cdot}77_0 . 10^{-10}$ e.s.u. ($4{\cdot}803_6$)

$e/c = 1{\cdot}5910_8 . 10^{-20}$ e.m.u. (1·6022)

Mass of electron:

$m_e = 9{\cdot}052 . 10^{-28}$ gm. (9·116)

Mass of hydrogen atom

$M_H = 1{\cdot}667_9 . 10^{-24}$ gm.

Ratio of mass of hydrogen atom to mass of electron:

$M_H/m_e = 1835{\cdot}6$

APPENDIX III

RYDBERG TERM TABLE

m	1	2	3	4	5	6	7	8	9	10	11	12	13
a +0·00	109737·10	27434·28	12193·01	6858·57	4389·48	3048·25	2239·53	1714·64	1354·78	1097·37	906·92	762·06	649·33
Δ	82302·82	15241·27	5334·44	2469·09	1341·23	808·72	524·89	359·86	257·41	190·45	144·86	112·73	
+0·05	99534·78	26112·34	11796·52	6690·27	4302·99	2998·08	2207·88	1693·41	1339·85	1086·48	898·73	755·75	644·37
Δ	73422·44	14315·82	5106·25	2387·28	1304·91	790·20	514·47	353·56	253·37	187·75	142·98	111·38	
+0·10	90691·82	24883·70	11419·05	6528·08	4219·04	2949·13	2176·89	1672·57	1325·17	1075·75	890·65	749·52	639·46
Δ	65808·12	13464·65	4890·97	2309·04	1269·91	772·24	504·32	347·40	249·42	185·10	141·13	110·06	
+0·15	82977·04	23739·77	11059·42	6371·73	4137·51	2901·37	2146·55	1652·11	1310·72	1065·18	882·68	743·36	634·60
Δ	59237·27	12680·35	4687·69	2234·22	1236·14	754·82	494·44	341·39	245·54	182·50	139·32	108·76	
+0·20	76206·33	22672·95	10716·51	6220·92	4058·33	2854·76	2116·84	1632·02	1296·52	1054·76	874·82	737·28	629·80
Δ	53533·38	11956·44	4495·59	2162·59	1203·57	737·92	484·82	335·50	241·76	179·94	137·54	107·48	
+0·25	70231·75	21676·47	10389·31	6075·41	3981·39	2809·27	2087·75	1612·30	1282·54	1044·49	867·06	731·28	625·06
Δ	48555·28	11287·16	4313·90	2094·02	1172·12	721·52	475·45	329·76	238·05	177·43	135·78	106·22	
+0·30	64933·18	20744·26	10076·87	5934·94	3906·63	2764·86	2059·24	1592·93	1268·78	1034·38	859·40	725·34	620·37
Δ	44188·92	10667·39	4141·93	2028·31	1141·77	705·62	466·31	324·15	234·40	174·98	134·06	104·97	
+0·35	60212·39	19870·91	9778·31	5799·29	3833·95	2721·48	2031·32	1573·91	1255·25	1024·41	851·85	719·48	615·73
Δ	40341·48	10092·60	3979·02	1965·34	1112·47	690·16	457·41	318·66	230·84	172·56	132·37	103·75	
+0·40	55988·32	19051·58	9492·83	5668·24	3763·28	2679·13	2003·96	1555·23	1241·93	1014·58	844·39	713·69	611·14
Δ	36936·74	9558·75	3824·59	1904·96	1084·15	675·17	448·73	313·30	227·35	170·19	130·70	102·55	
+0·45	52193·63	18281·90	9219·67	5541·58	3694·54	2637·75	1977·16	1536·88	1228·82	1004·90	837·03	707·97	606·61
Δ	33911·73	9062·23	3678·09	1847·04	1056·79	660·59	440·28	308·06	223·92	167·87	129·06	101·36	
+0·50	48772·04	17557·94	8958·13	5419·12	3627·67	2597·33	1950·88	1518·85	1215·92	995·35	829·77	702·32	602·12
Δ	31214·10	8599·81	3539·01	1791·45	1030·34	646·45	432·03	302·93	220·57	165·58	127·45	100·20	

m	1	2	3	4	5	6	7	8	9	10	11	12	13
a +0·55	45676·21	16876·14	8707·56	5300·67	3562·60	2557·83	1925·13	1501·14	1203·22	985·94	822·60	696·73	597·69
Δ	28800·07	8168·58	3406·89	1738·07	1004·77	632·70	423·99	297·92	217·28	163·34	125·87	99·04	
+0·60	42866·05	16233·30	8467·37	5186·06	3499·27	2519·22	1899·88	1483·74	1190·72	976·66	815·53	691·21	593·30
Δ	26632·75	7765·93	3281·31	1686·79	980·05	619·34	416·14	293·02	214·06	161·13	124·32	97·91	
+0·65	40307·47	15626·50	8236·97	5075·13	3437·61	2481·48	1875·13	1466·63	1178·42	967·51	808·54	685·76	588·96
Δ	24680·97	7389·53	3161·84	1637·52	956·13	606·35	408·50	288·21	210·91	158·97	122·78	96·80	
+0·70	37971·32	15053·10	8015·86	4967·73	3377·57	2444·58	1850·85	1449·82	1166·30	958·49	801·64	680·37	584·67
Δ	22918·22	7037·24	3048·13	1590·16	932·99	593·73	401·03	283·52	207·81	156·85	121·27	95·70	
+0·75	35832·53	14510·69	7803·53	4863·69	3319·08	2408·50	1827·05	1433·30	1154·37	949·59	794·84	675·05	580·43
Δ	21321·84	6707·16	2939·84	1544·61	910·58	581·45	393·75	278·93	204·78	154·75	119·79	94·62	
+0·80	33869·48	13997·08	7599·52	4762·89	3262·10	2373·21	1803·70	1417·06	1142·62	940·82	788·12	669·78	576·23
Δ	19872·40	6397·56	2836·63	1500·79	888·89	569·51	386·64	274·44	201·80	152·70	118·34	93·55	
+0·85	32063·44	13510·26	7403·42	4665·20	3206·58	2338·69	1780·80	1401·09	1131·05	932·17	781·48	664·58	572·08
Δ	18553·18	6106·84	2738·22	1458·62	867·89	557·89	379·71	270·04	198·88	150·69	116·90	92·50	
+0·90	30398·09	13048·41	7214·80	4570·47	3152·46	2304·92	1758·33	1385·39	1119·65	923·63	774·93	659·44	567·97
Δ	17349·68	5833·61	2644·33	1418·01	847·54	546·59	372·94	265·74	196·02	148·70	115·49	91·47	
+0·95	28859·20	12609·84	7033·30	4478·61	3099·70	2271·87	1736·28	1369·96	1108·43	915·22	768·45	654·36	563·90
Δ	16249·36	5576·54	2554·69	1378·91	827·83	535·59	366·32	261·53	193·21	146·77	114·09	90·46	
+1·00	27434·28	12193·01	6858·57	4389·48	3048·25	2239·53	1714·65	1354·78	1097·37	906·92	762·06	649·33	559·88
Δ	15241·27	5334·44	2469·09	1341·23	808·72	524·88	359·87	257·41	190·45	144·86	112·73	89·45	

$m+a=$	0·95	0·90	0·85	0·80	0·75	0·70	0·65	0·60	0·50
	121592·4	135477·9	151885·3	171464·2	195088·2	223953·4	259732·8	304825·9	438948·4

APPENDIX IV

NOTATION

The notation used in the description of line spectra should not conflict with that used in the description of molecular spectra; this is axiomatic. I hope I have paid due attention to the notation adopted by Jevons in his Report on Band Spectra.

λ: wave-length *in vacuo*, expressed in Angstrom units.
A.: Angstrom unit, 10^{-8} cm.
X: X unit, 10^{-3} A.
c: velocity of light *in vacuo*.
ν': frequency of vibration *in vacuo*. $\nu' = c/\lambda$.
ν: wave-number in cm.$^{-1}$ $\nu = 1/\lambda$.
E: the energy of an atom in ergs measured from the ground state of the atom.
E_ν: the energy of an atom in cm.$^{-1}$;—strictly this should not be called 'energy', but one can be too pedantic.
$E^\star$: the energy of an atom in ergs, measured from the ground state of the ion.
W: the energy in ergs, which must be added to an atom to remove an electron to infinity. $W = -E^\star$.
T: term value in cm.$^{-1}$ measured from the limit of a term sequence. $T = W/ch$.
λ_l: wave-length of a series limit.
T_l: wave-number of a series limit. $T_l = 1/\lambda_l$.
$'$ and $''$: superscripts affixed to symbols which refer to the upper and lower states respectively; thus $\nu = T'' - T'$.
$\rightarrow$ and $\leftarrow$: denote emission and absorption; symbols for the upper and lower levels are always written to the left and right respectively of the arrows. A commoner notation uses a minus sign and places the lower level to the left.
R and $n^\star$: Rydberg constant and the effective quantum number. $T = R/n^{\star 2}$.
m: the serial number.
a: the quantum defect, defined by $n^\star = (m+a)$, where m is integral and a usually less than 1.
Na I, Na II, Na III, ...: the arc, spark and second spark spectra of sodium. There seems good reason for adopting the suggestion made at a recent meeting of the Physical Society and writing these spectra Na, Na^+ and Na^{2+}, thus conforming to the notation of band spectra; if HCl I is awkward, so also is I I. But I have hesitated to break a convention, which is universal.
n: the chief quantum number.
$\mathbf{l}$ and $\mathbf{s}$: the orbital and spin angular momenta of a single electron.
$\mathbf{j}$: angular momentum of an electron moving in an orbit. $\mathbf{j} = \mathbf{l} + \mathbf{s}$.
l, s, j: quantum numbers which determine $\mathbf{l}$, $\mathbf{s}$ and $\mathbf{j}$. $\mathbf{l} = l.h/2\pi$.
$\mathbf{L}$: atomic orbital momentum. $\mathbf{L} = \mathbf{l}_1 + \mathbf{l}_2 + \ldots$.
$\mathbf{S}$: atomic spin momentum. $\mathbf{S} = \mathbf{s}_1 + \mathbf{s}_2 + \ldots$.

J: electronic angular momentum; the resultant angular momentum of a number of electrons. $\mathbf{J}=\mathbf{L}+\mathbf{S}=\mathbf{j}_1+\mathbf{j}_2+\ldots$

I: spin of the nucleus. Jevons uses T for this.

F: angular momentum of the complete atom. $\mathbf{F}=\mathbf{J}+\mathbf{I}$.

L, S, J, I, F: quantum numbers which determine the angular momenta **L, S, J, I, F**. $\mathbf{L}=L.h/2\pi$.

s, p, d: states of electron in which $l=0, 1, 2, \ldots$

S, P, D: states of atom in which $L=0, 1, 2, \ldots$

H: intensity of a magnetic field in gauss.

F: intensity of an electric field in volt/cm.

m, m_s, m_l: quantum numbers which determine the projections of **j**, **s** and **l** on the axis of a magnetic or electric field. $m=j\cos(\mathbf{jH})$.

M, M_S, M_L: quantum numbers which determine the projections of **J**, **S** and **L** on the axis of a magnetic or electric field. $M=J\cos(\mathbf{JH})$.

π and σ components: components polarised parallel or perpendicular to the magnetic or electric field.

m_e: mass of an electron.

e: charge on an electron in electrostatic units.

M: mass of the nucleus.

Z: atomic number.

o_m: Lorentz unit or normal Zeeman interval in cm.^{-1}

g: Landé's magnetic splitting factor. $\Delta E/ch=go_m$.

μ: the magnetic moment of an atom or a nucleus in erg gauss^{-1}.

μ_B: Bohr magneton.

μ_N: nuclear magneton. $\mu_N=\mu_B/1838$.

ν_G: the centroid of a multiplet in units cm.^{-1}

Γ: the displacement of a component from the centroid of a multiplet term. $\Delta E/ch=\nu_G+\Gamma$.

γ: electronic displacement.

A: atomic interval quotient, or coefficient of magnetic interaction of resultant orbital and spin angular momenta.

a: electronic interval quotient, or coefficient of magnetic interaction of electronic orbital and spin angular momenta.

SUBJECT INDEX

Numbers in Clarendon refer to the second volume.

AUTHOR INDEX

Numbers in Clarendon refer to the second volume.

Printed in the United States
By Bookmasters